T3-BSF-315

Skull Morphology of *Lambdopsalis bulla* (Mammalia, Multituberculata) and its Implications to Mammalian Evolution

Frontispiece

Restoration of a Chinese late Paleocene taeniolabidid multituberculate mammal, *Lambdopsalis bulla* (drawn by Claire Vanderslice).

Special Paper 4

Miao (Desui)

Skull Morphology of *Lambdopsalis bulla* (Mammalia, Multituberculata) and its Implications to Mammalian Evolution

Edited by
Donald W. Boyd
Jason A. Lillegraven

Contributions to Geology
The University of Wyoming

Preferred mode of citation:

Miao, D., 1988, Skull morphology of *Lambdopsalis bulla* (Mammalia, Multituberculata) and its implications to mammalian evolution: Contributions to Geology, University of Wyoming, Special Paper 4, *viii* + 104 p.

Department of Geology and Geophysics
The University of Wyoming
Laramie, Wyoming U.S.A. 82071-3006

ISBN 0-941570-09-6
Library of Congress Catalog Card Number: 88-51282
Printed in the United States of America

DEDICATION

This contribution is dedicated to:

Chow Minchen and Zhai Renjie, my first teachers in vertebrate paleontology, whose fondness and support are greatly appreciated.

Bill Clemens, who helped me sweepstake across the world's largest patch of water to pursue my study on its east side, and told me: "Once upon a time, there lived enigmatic beasts called multituberculates. . ."

Jay Lillegraven, my mentor, cheerleader, and tolerator (as well as nitpicking editor) throughout the writing of the following pages.

QUOTATIONS

Without faith a man can do nothing. But faith can stifle all science.
Henry F. Amiel

Innocent, unbiased observation is a myth.
Peter B. Medawar

The business of scientists is not to corroborate previous beliefs but to test them.
Certainty is boring—and not really achievable.
Malcolm C. McKenna

CONTENTS

Skull morphology of *Lambdopsalis bulla* (Mammalia, Multituberculata) and its implications to mammalian evolution

MIAO DESUI

*Department of Geology and Geophysics, The University of Wyoming, Laramie, Wyoming 82071-3006**

ABSTRACT

Multituberculates are an extinct mammalian order that lived in Mesozoic and early Cenozoic eras. *Lambdopsalis*, a Paleocene multituberculate recovered in China, preserves cranial remains that allow in this study: (1) a description of its skull morphology; (2) a reconstruction of its nonfossilized structures such as the cranial nerve system and major cranial vasculature; (3) an analysis of functional adaptation of its auditory system; and (4) an interpretation of phylogenetic relationships within multituberculates themselves and among other major mammalian groups.

Character analysis reveals that a number of previously used cranial features in reconstructions of mammalian phylogeny are unreliable. These include premaxillo-frontal contact, exclusion of septomaxilla from face, number of infraorbital foramina, extent of orbital exposure of palatine, presence versus absence of jugal, lacrimal, and parasphenoid, extent of cranial process of squamosal, and reduction of alisphenoid.

The bulla-like structure of *Lambdopsalis* is the expanded vestibular apparatus, not an enlarged tympanic bulla. The expanded vestibular apparatus, flat incudomalleal joint, and absence of a well defined fossa muscularis minor in *Lambdopsalis* suggest that *Lambdopsalis* (possibly a burrower) adapted to low-frequency perception.

Lambdopsalis possesses a large alisphenoid (perforated by the trigeminal foramina) and a slender "anterior lamina of the petrosal." The discovery supports Presley's (1981) argument of close affinity between "nontherian" and "therian" mammals, and invalidates the hypothesis of fundamental nontherian/therian dichotomy. Contrary to general consensus, available paleontological evidence does not indicate the existence of a uniform structural pattern of the braincase for nontherian groups.

Cranial characters coupled with dental features document monophyly for nonharamiyid multituberculates. The skull morphology of *Lambdopsalis* shows phylogenetic unity with taeniolabidids, and invalidity of Lambdopsalidae Chow and Qi, 1978.

Assuming monophyly of Mammalia, the class is divisible into a crown group and a stem group. The crown group includes all living mammals plus the fossil therians that shared the latest common ancestor with monotremes. The stem group consists of all remaining extinct mammals. Multituberculates belong to the paraphyletic stem group, and diverged from the main lineage leading to living mammals prior to emergence of the latest common ancestor of modern mammals. More intimate relations among members of the stem group remain uncertain, but are limited to but a few reasonable alternatives.

SECTION I

INTRODUCTION

Multituberculates are an extinct order of early mammals whose temporal range spans at least from the Late Jurassic to early Oligocene. However, the inclusion of haramiyids in the order Multituberculata (see Hahn, 1973) would further extend its geologic history back to the Late Triassic. They are, therefore, so far known to be the longest lived order in mammalian history. To date, they have an exclusively Holarctic distribution. Bonaparte (1986) reported discovery of some multituberculate teeth from the Cretaceous of Argentina. This would expand multituberculate distribution to a southern continent, if the teeth are correctly identified. Multituberculates are presumed to have been ecological counterparts of rodents before rodents evolved, and were extremely diverse. Their extinction has been attributed to competitive exclusion by contemporary herbivorous placentals in similar niches (Van Valen and Sloan, 1966; Krause, 1986). Half a century ago, Simpson (1937, p. 727) introduced the order in this way:

> "The order Multituberculata is of interest and importance as having been a dominant mammalian group throughout the Mesozoic, as covering a larger span than any other order of

*Present address: Department of Anatomy, The University of Chicago, Chicago, Illinois 60637.

mammals, as having a peculiar and puzzling anatomical structure, and as bearing on numerous essential problems of modes and methods of evolution, of mammalian classification, of molar history, and many others."

Although multituberculate teeth and jaw fragments are fairly common in many adequately sampled Late Cretaceous and Paleogene Holarctic local faunas, cranial materials usually are rare. The oldest known cranial materials, found in Kimmeridgian beds of Portugal (Hahn, 1969), consist of a crushed skull of *Paulchoffatia* and a rostrum of *Kuehneodon*. Hahn's reconstruction (*ibid.*) of the skull of *Paulchoffatia* was based essentially on Simpson's reconstruction (1937) of the skull of *Ptilodus*, especially for the otic region and the part of the basicranium. Therefore, it includes some speculative elements. With additional findings of fragmental paulchoffatiid crania, Hahn (1977, 1981) made further efforts to reconstruct their skull structures, notably adding more information to reconstructions of the snouts of *Kuehneodon* and *Henkelodon*, of the orbital region of *Henkelodon*, and of the sphenoid region of *Pseudobolodon*. However, again due to the poor preservation of the materials, these reconstructions seem to be more interpretive than descriptive.

Simpson (1925) published the first discovery of Late Cretaceous multituberculate cranial material, an anterior part of the skull of *Djadochtatherium matthewi*, collected in Mongolia during the Third Asiatic Expedition of the American Museum of Natural History. Though the preservation is poor, Simpson's reconstruction of the anterior half of the skull is reasonable. For the taxonomic status of *Djadochtatherium*, see discussion by Simmons and Miao (1986).

Kielan-Jaworowska (1970, 1971, and 1974) described a rich collection of the Mongolian Late Cretaceous multituberculate skulls collected by the Polish-Mongolian Paleontological Expeditions. Reconstructions of these skulls were summarized by Clemens and Kielan-Jaworowska (1979), and became a standard and authoritative source for information on multituberculate cranial anatomy. Due to the then prevalent ideas of nontherian versus therian dichotomy, Kielan-Jaworowska interpreted the structure of the lateral wall of the braincase of the Mongolian multituberculates as being similar to that of *Ornithorhynchus*, one of the living representatives of so-called nontherians (or "prototherians"). She also interpreted the structure of the occipital region of those Mongolian multituberculates as being more reptilian than they actually are. Kielan-Jaworowska also criticized Simpson's reconstruction of the skull of *Ptilodus* as being hypothetical, developed on essentially therian lines. Subsequently, however, she and her co-workers gradually have modified these interpretations in recent studies of serially sectioned skulls from the aforementioned collection (Kielan-Jaworowska and others, 1984; 1986 and references therein).

The earliest studied multituberculate skulls, however, are those from the North American Paleocene. Gidley (1909) first described a nearly complete skull of *Ptilodus* from the Fort Union Formation of Montana. Five years later, Broom (1914) described a skull of *Taeniolabis* (then called *Polymastodon*) from the "Puerco" Formation (=Nacimiento Formation) of New Mexico. The same specimen was reinvestigated subsequently by Granger and Simpson (1929) and Simpson (1937). With additional findings of the cranial materials of *Ptilodus*, Simpson (1937) also made a fine effort in reconstructing their structure. Because of poor preservation, however, it has been impossible to recognize details of cranial anatomy of *Taeniolabis*. But Simpson did manage to determine some of the structural details of the braincase of *Ptilodus*, especially the basicranium, with great credibility.

The youngest known multituberculate skull is that of *Ectypodus tardus*, an early Eocene ptilodontoid from Wyoming. Although Sloan (1979) provided a figured reconstruction of the skull, he gave no actual description of it. The anatomical features shown in the figure are an interesting melange of those reconstructed by Simpson (1937) and Kielan-Jaworowska (1971, 1974) plus those of modern therians. For example, it shows an extended anterior lamina of the petrosal and, yet, a well-developed alisphenoid with the foramina rotundum and ovale passing through the alisphenoid. It also shows a bony tympanic ring (see also Allin, 1986). However, some of Sloan's general discussions are questionable. For example, he stated (p. 493): "The cochlea of multituberculates is straight with a tiny hook at the end much as in cynodonts, and is about one sixth the length of that of Late Cretaceous-Eocene placentals of similar size." As far as known, neither multituberculates nor cynodonts have been described as having a cochlea with a hook at the end; instead, their cochleae are straight. Multituberculates vary considerably in size among different taxa (*e. g.*, *Ectypodus* versus *Taeniolabis*), and so do their cochleae. The same is true of "Late Cretaceous-Eocene placentals." Sloan should have specified which taxa permitted this quantitative comparison. The lack of adequate description to accompany the gross discussion has inevitably led to doubts of the validity of Sloan's figured reconstruction.

Interesting additions to this list of multituberculate cranial materials are well-preserved skulls of *Lambdopsalis bulla* Chow and Qi, 1978 from the latest Paleocene or earliest Eocene of Inner Mongolia, People's Republic of China. *Lambdopsalis* was first found by Chow Minchen and his field team in 1975 from its type locality in the Nomogen area, Siziwang Qi County, Inner Mongolia (according to Li and Ting [1983], Nomogen formerly was known as Nom Khong Shireh or Nomogen Ora, see also Radinsky [1964]). Chow and others (1976) first announced the finding in a short report basically dealing with stratigraphy of the area. Later, Chow and Qi (1978) published a paper containing a faunal list of the area, with brief accounts of its major taxa. In the same paper, they named and described *Lambdopsalis bulla*.

Soon after publication, the taxonomic validity of *Lambdopsalis* was challenged by Kielan-Jaworowska and Sloan (1979). Among a number of other criticisms, they thought *Lambdopsalis* to be very close to *Prionessus*, and possibly a junior synonym of *Sphenopsalis*, and therefore believed it does not merit a new generic name. The challenge necessitated a more careful study of *Lambdopsalis*.

Besides the material of *Lambdopsalis* described by Chow and Qi, Zhai Renjie collected more and better preserved specimens of the same taxon from another locality, the type section of the Bayan Ulan Formation. This locality is situated in the northern foot of Nomogen Mesa, about 60 km north of the type locality in Nomogen (for further information about the localities, see Miao, 1986).

These excellent materials attracted attention of William A. Clemens and Malcolm C. McKenna when they visited the Institute of Vertebrate Paleontology and Paleoanthropology (IVPP) in Beijing in the summer of 1981. They suggested this project, and I was honored to be chosen among the first group of graduate students at IVPP after the Cultural Revolution to take over the study. Chow, Zhai, and Qi generously lent all the materials. The first part of this project (Miao, 1986) was completed at the Museum of Paleontology, University of California, Berkeley. It documented the taxonomic validity of *Lambdopsalis* and phylogenetic unity of the Asian Tertiary multituberculates *Prionessus*, *Sphenopsalis*, and *Lambdopsalis*. A second part was the report of discovery of the three ear ossicles in *Lambdopsalis* (Miao and Lillegraven, 1986). The present paper is the third part of the project. A planned fourth part will be devoted to postcranial anatomy of *Lambdopsalis*.

The aims of the present paper are: (1) to describe skull morphology of *Lambdopsalis* and make comparisons with that of other multituberculates and other groups of early mammals; (2) to reconstruct osteological patterns of the skull; (3) to interpret its nonfossilized structures such as the cranial nerve system and major drainages of cranial vasculature; (4) to analyze the functional adaptation of its auditory system; and (5) to discuss phylogenetic relationships within multituberculates themselves and among other early mammalian groups.

Unless specified otherwise, anatomical terminology used in the present paper follows the usages by McDowell (1958) and Kielan-Jaworowska and others (1986) for the skull in general, and that of MacPhee (1981) for the ear region in particular. *Lambdopsalis* and many other fossil taxa mentioned in the paper are monotypic and, for brevity, will be referred to only by their generic names.

It also should be noted that the materials used in this study include about a dozen skulls, hundreds of jaw fragments and isolated petrosals, and many postcranial elements. They were recovered by half a dozen collectors within only a few hours of surface picking. The fossils were weathered out from yellowish or reddish claystones and siltstones. The bone is hard and the matrix is reasonably soft, except for certain local calcifications. All preparation except that of the type specimen (IVPP V5429) was done by me with the aid of tungsten carbide needles. Specimens with prefix number V5429 are from the type locality of *Lambdopsalis* (at the type section of the Nomogen Formation), and collected mainly by Qi Tao. Specimens with prefix numbers V7151 and V7152 are from the type section of the Bayan Ulan Formation, and collected mainly by Zhai Renjie and also by the Clemens and the McKennas.

In discussion of correlation of mammal-bearing Late Cretaceous formations of Mongolia, Lillegraven and McKenna (1986) pointed out: "Attempts to determine ages of faunas recovered from these Asian Late Cretaceous continental sediments have not had the benefit of radioisotopic or magnetostratigraphic data and have involved biostratigraphic methods alone . . ." This comment is equally applicable to the age designation of *Lambdopsalis bulla*, whose type specimen is from the Nomogen Formation, presumably deposited during the late Paleocene. But the majority of specimens under study is from the Bayan Ulan Formation, postulated as representing the early Eocene (Li and Ting, 1983). These age determinations were based solely on the known geologic distribution of other members of the associated faunas (for faunal lists of these two formations see Li and Ting, 1983). On the same biostratigraphic basis, however, Zhai argued the Bayan Ulan Formation as well as the Nomogen Formation might be early Eocene in age. Clearly, solution of the dispute lies in further refinement of the regional stratigraphy with more sophisticated geologic dating techniques.

ABBREVIATIONS

Specimen Institutional Identifications

- AMNH — Department of Vertebrate Paleontology, The American Museum of Natural History, New York
- BM — The British Museum (Natural History), London
- CUP — Catholic University of Peking
- IVPP — Institute of Vertebrate Paleontology and Paleoanthropology, The Chinese Academy of Sciences, Beijing
- USNM — National Museum of Natural History, Smithsonian Institution, Washington, D.C.

Tooth Abbreviations

More than ever before, we now appreciate the convenience of using the upper versus lower case letters to stand for upper versus lower teeth, respectively, as the late Professor Jepsen suggested many years ago. I adopt this way of abbreviation in the present paper, for example:

- M1 — First upper molar
- m1 — First lower molar

Anatomical Abbreviations

- aas — ampulla of anterior semicircular canal
- ac — ascending canal
- acf — ascending canal foramen
- adm — arteria diploetica magna
- afpc — anterior foramen of post-temporal canal
- al — alisphenoid

alcf alisphenoid canal foramen
alp anterior lamina of petrosal
als ampulla of lateral semicircular canal
aps ampulla of posterior semicircular canal
aqc aquaeductus cochleae
aqF aquaeductus Fallopii
aqv aquaeductus vestibuli
asc anterior semicircular canal

bo basioccipital
bs basisphenoid

cc cochlear cavity
ce cavum epiptericum
cfaF cranial foramen of aquaeductus Fallopii
cfpc cranial foramen of prootic canal
cma canal of maxillary artery (= ramus inferior herein)
crc crus commune
csc cavum supracochleare

ec endocast of cochlea
ef ethmoid foramen
er epitympanic recess
es endolymphatic sac
evf emissary venous foramina

f frontal
fc fenestra cochleae
fcn foramen for cochlear nerve
ffp foramen facialis primitivum
ffs foramen facialis secondarium
fm foramen masticatorium
fmm fossa muscularis major
foi foramen ovale inferium
fs facial sulcus
fv fenestra vestibuli
fvn foramen for vestibular nerve

gf glenoid fossa
gps groove for prootic sinus
gss groove for sigmoid sinus

hgf hypoglossal foramen
hmp hyoid muscle pit

if incisive foramen
iff infraorbital foramen
ijv internal jugular vein
ioc infraorbital canal

jf jugular foramen

lr lateral ridge
lsc lateral semicircular canal

m maxilla
mg median gutter
mgr median groove
mipf minor palatine foramen
mjpf major palatine foramen
mof metoptica foramen

n nasal
nlf nasolacrimal foramen

obr oblique ridge
oc occiptial
ocl occipital condyle
onf orbitonasal fossa
opf optic foramen
or orbital ridge
os orbitosphenoid

p parietal
pas parasphenoid
pe petrosal
pfc prefacial commissure
pgf postglenoid foramen
pl palatine
pm premaxilla
pof postorbital foramen
pop postorbital process
potc post-temporal canal
pp paroccipital process
pr promontorium of petrosal
prc prootic canal
prcf prootic canal foramen
prf prootic foramen
prr promontorium recess
prs presphenoid
psc posterior semicircular canal
pt pterygoid
ptcf pterygoid canal foramen
ptf post-temporal fossa
pv prootic vein (sinus)

res recess of endolymphatic sac
ri ramus inferior of stapedial artery (= "maxillary artery" *sensu* Kielan-Jaworowska and others, 1986)
rio ramus infraorbitalis
rs ramus superior (= "ascending canal vessel")
rso ramus supraorbitalis

sf subarcuate fossa
sgf supraglenoid foramen
sgv supraglenoid vein
smn stylomastoid notch
smv stylomastoid vein
sof sphenorbital fissure
spf sphenopalatine foramen
sq squamosal
ss sigmoid sinus
st sella turcica
sta stapedial artery
stp styloid process

tc transverse crest
tr transverse ridge
ts transverse sinus

v vomer
vcl vena capita lateralis (= lateral head vein)

II optic nerve
V1 ophthalmic branch of trigeminal nerve
V2 maxillary branch of trigeminal nerve
V3 mandibular branch of trigeminal nerve
VII facial nerve

SECTION II

ORGANIZATION

Anatomical descriptions of cranial bones have been organized by various authors according to one of the following criteria: (1) functional; (2) developmental; (3) topographical; and (4) any combination of these. For convenience as well as personal preference, I choose a combination of developmental and topographical criteria in presenting morphological information in this paper. The cranial bones will be treated as follows: (1) dermal skull roof; (2) palatal complex; (3) braincase; and (4) lower jaw.

Under subtitle of each individual bone being considered, two subdivisions of the section are generally given: (1) *Description*, or "hard evidence," to include only objective anatomical observations, repeatable by other researchers; and (2) *Discussion* including morphological comparisons as well as developmental, functional, and phylogenetic interpretations or speculations (which, to various readers, may be debatable). For example, interpretations of soft anatomy from osseous evidence in fossils, as Novacek (1980) pointed out, probably can never be fully justified. Conroy and Wible (1978) found that the canal usually identified as carrying the promontory artery conveys instead the internal carotid nerve in *Lemur variegatus*. Gaughran (1954) showed that no palatine branch (nasopalatine nerve) from the sphenopalatine ganglion can be traced to the incisive foramen in *Blarina brevicauda*, which is otherwise a consistent feature among mammals. Thus, exceptions of this sort should weaken somewhat our confidence (but probably should not discourage our effort) in inferring nonfossilized structures from fossilized osseous ones. Nevertheless, the close association between soft parts and osseous structures in a wide range of modern representatives points to a high degree of credibility of the interpretations applied to fossil forms.

A clear separation of observation from interpretation is rendered crucially important, based both upon the exceptions mentioned above and upon my acceptance of what Darwin (1896) once had to say:

"False facts are highly injurious to the progress of science, for they often endure long; but false views, if supported by some evidence, do little harm, for everyone takes a salutary pleasure in proving their falseness; and when this is done, one path towards error is closed and the road to truth is often at the same time opened."

DERMAL SKULL ROOF

Premaxilla (Figs. 1-4, 12, 17, 18)

Description

The premaxillary bone can be well observed on the specimens IVPP V5429, V5429.1-2, V5429.10, V7151.1, V7151.50, V7151.55, V7151.79, V7151.82-83, and V7151-97. It consists of a horizontal palatal process, an internarial process, and an extensive, more-or-less perpendicular nasal process. It bears a larger I2 and a smaller I3 at its labial edge, and forms walls of the anterior part of the nasal cavity.

The premaxillo-maxillary suture is interdigitated and the premaxilla is overlapped by the maxilla. The premaxillo-maxillary suture in palatal view is perpendicular to the anteroposterior axis of the skull, and situated at about the half-way point between the anterior margin of the snout and P4. This suture in lateral view is obliquely extended posterodorsad. It meets the nasal bone at the posterior two-fifths of the nasal.

The dorsal parts of the internarial processes wedge between the anterior ends of the nasals. Although an internarial bar is not preserved in any available specimen, the suture between the internarial process of the premaxilla and the nasal is clearly visible in dorsal view of IVPP V5429.10, V7151.50, and V7151.82, and in ventral view of V7151.97. Bone is thickened along the suture zone between the posterodorsal parts of the internarial processes; the suture between the internarial processes disappeared at least in V7151.97.

The palatal processes of the premaxillae are undisruptedly sutured medially, and form the anterior part of the floor of the nasal cavity. Paired incisive foramina are present on the palatal processes. They are shaped as anteroposteriorly elongated ovals. Their anterior edges are at the posterior margins of the I2s, with posterior edges close to and anterior to the premaxillo-maxillary suture. Medially, the foramina are bordered by thin crests of the suture zone of the palatal processes. Immediately anteromedial to the incisive foramina, an anteroposteriorly-elongated oval depression is developed along the suture zone between the palatal processes.

Discussion

The premaxilla in *Lambdopsalis* is a dermal bone (sense of C. Patterson, 1977) that bears upper incisors and forms the internarial bar and the walls of the anterior part of the nasal cavity. However, all previous authors reconstructed the external nares in multituberculates as confluent, as is characteristic of all living mammals. Among known early mammals, only *Sinoconodon* is considered as having an internarial bridge (Patterson and Olson, 1961; Crompton and Sun, 1985). In their description of *Morganucodon*, Kermack and others (1981, p. 2) stated: "In no specimen is there a trace of a median dorsal process. It is, therefore, most probably that the external nares were confluent as in tritylodonts and—according to our interpretation—*Sinoconodon rigneyi*." It should be noted, however, that: (1) the external nares in tritylodonts, with only the possible exception of *Oligokyphus*, are not confluent (Young, 1947; Kühne, 1956; Clark and Hopson, 1985; and Sues, 1986); (2) the specimen of *Sinoconodon rigneyi* has been further prepared since Kermack and others' writing, and it appears rather suggestive of the presence of an internarial bar (*personal observation*); and (3) although, as the authors

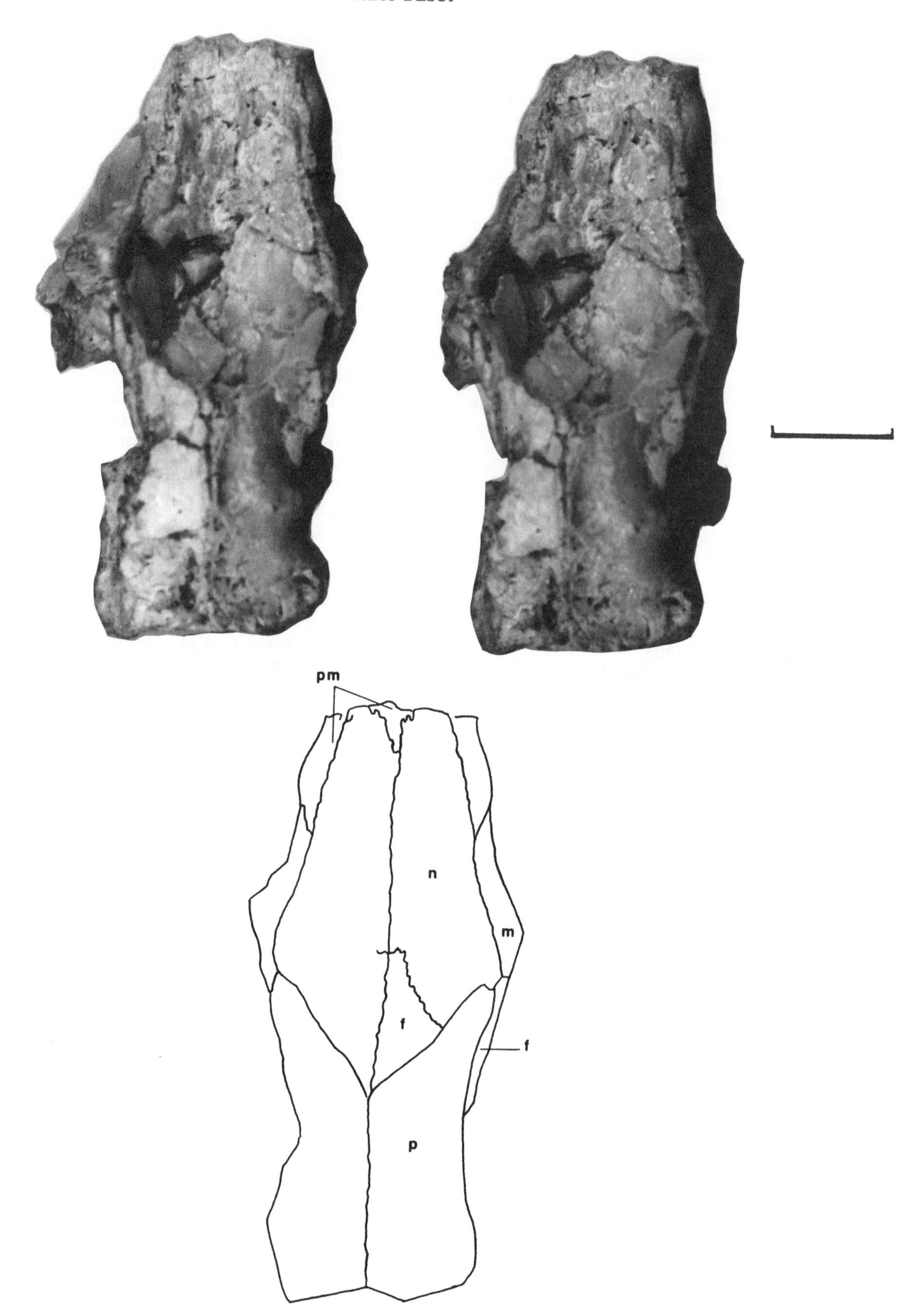

Figure 1. *Lambdopsalis bulla*, V7151.50, dorsal stereoscopic view of skull. Anterior to top. Scale bar in this, and all other illustrations of specimens, is one centimeter and refers only to accompanying photograph(s).

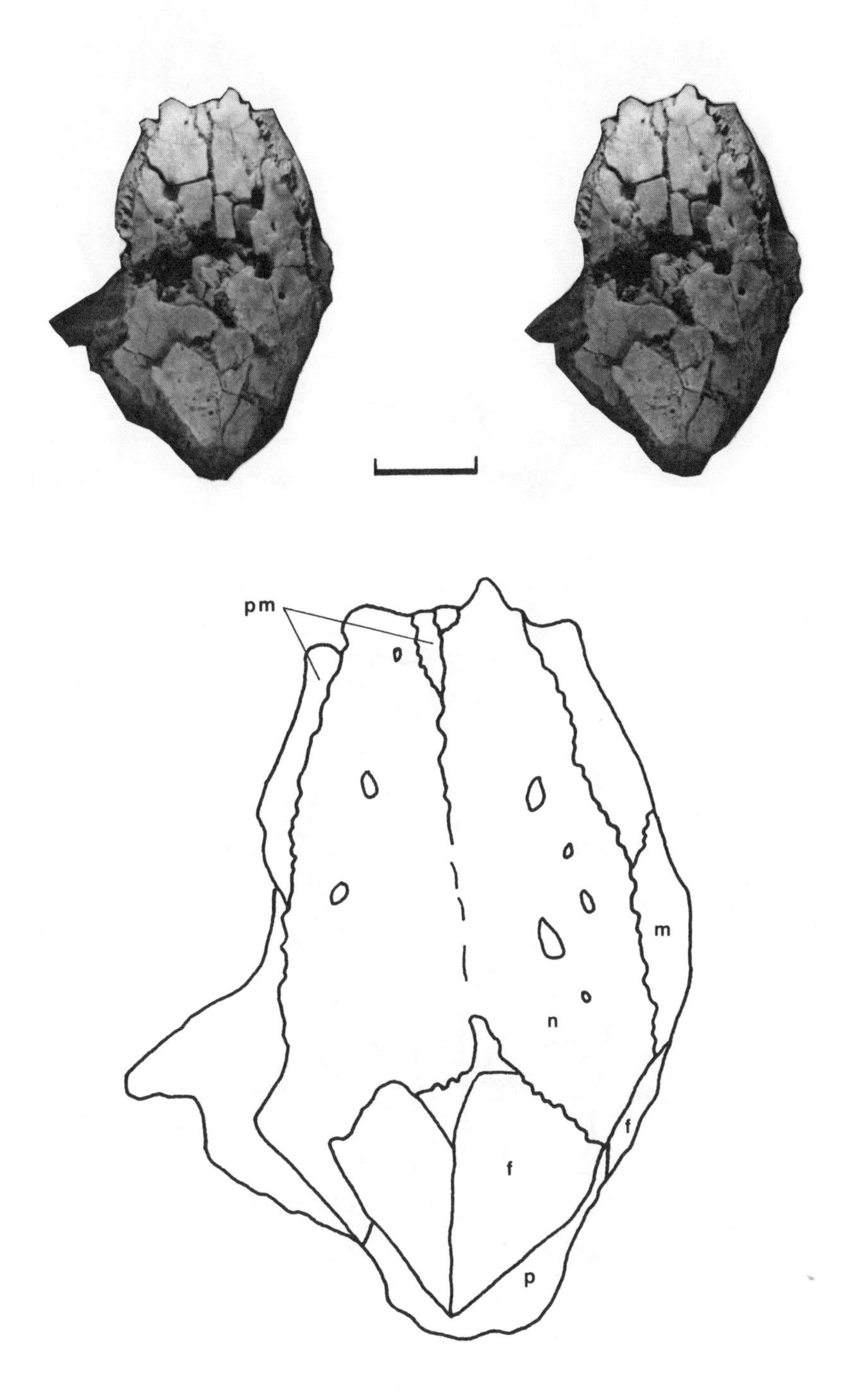

Figure 2. ***Lambdopsalis bulla*, V5429.10, stereophotograph, dorsal view of fragment of skull showing dorsal part of internarial process of premaxilla. Anterior to top.**

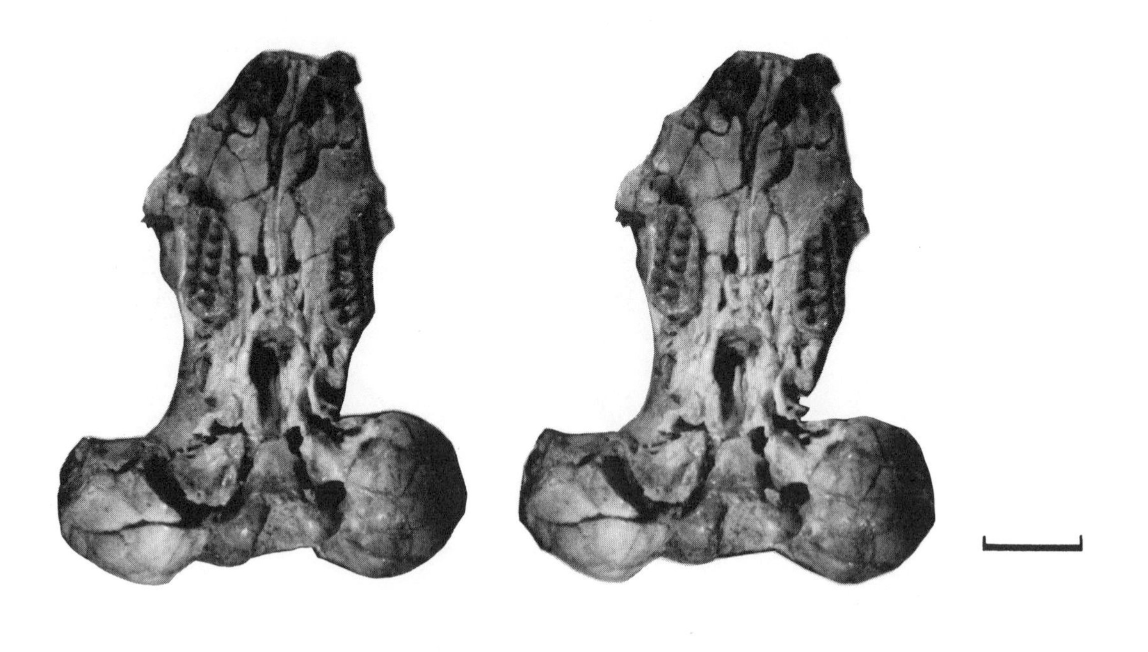

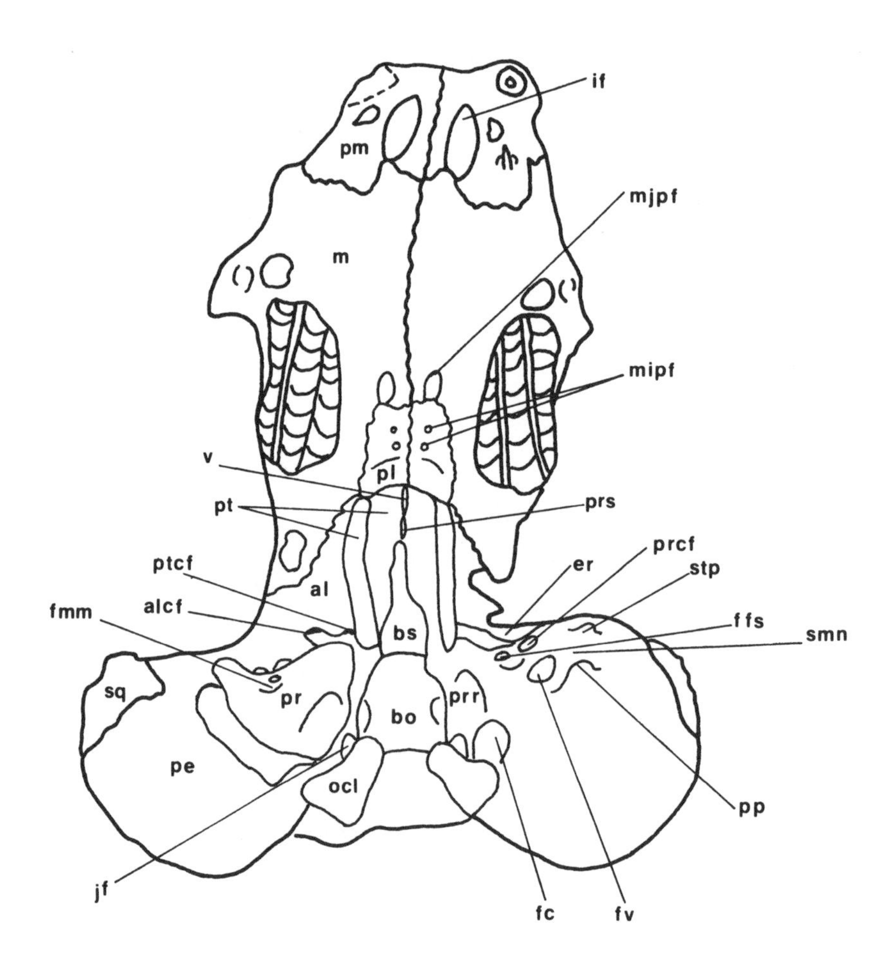

Figure 3. ***Lambdopsalis bulla*****, V 5429, holotype; ventral stereoscopic view of skull. Anterior to top.**

claimed in the above-quoted lines, it seems no trace of a median dorsal process of the premaxilla is preserved in the specimens of *Morganucodon* available to them, it is also true that the anterior dorsal roof of the nasal cavity (a crucial area which would show whether there is trace of the median dorsal process) is totally gone. In fact, Kermack and others (1981) reconstructed the length of the nasal on the basis of impressions remaining on the poorly preserved endonasal cast of *M. oehleri*, CUP 2320 (see their Figs. 2, 3, and 95). Contrariwise, their Figures 95, 99, 100, and 102 all show hints of probable presence of an internarial process of the premaxilla. Although broken, an elevated process at the anterior end of the palatal process of the premaxilla is shown clearly in those figures.

An undescribed specimen (mam 41/75) of the docodont *Haldanodon* (housed in West Berlin's Institute of Paleontology, Free University) also shows the presence of an internarial process of the premaxilla (*personal observation*). Thus, it seems conceivable that an internarial bar remains throughout various groups of early mammals. The reported absence of an internarial bar in other multituberculates might be more apparent than real, especially when one considers: (1) the usual poor preservation of the most anterior part of the snout; and (2) that insufficient attention has been paid to anatomy of this particular area. Be that as it may, the loss of an internarial bar potentially can be treated as a synapomorphy shared by monotremes and therians.

Alternatively, if one assumes that an internarial bar were indeed lost in other, especially more primitive, multituberculates, one has to interpret the presence of the internarial bar in *Lambdopsalis* as an evolutionary reversal. Although not impossible, it seems improbable in light of distribution of the character in diverse groups of early mammals. In addition, the stereophotograph of Figure 16 of Kielan-Jaworowska and others (1986) appears indicative of the presence of a sutural facet at the anterior end of the left nasal in *Chulsanbaatar vulgaris,* ZPAL Mg M-I/89. If this proves to be the case, the bone anterior to the nasal must be the dorsal part of the internarial process of the premaxilla. Therefore, this warrants reexamination of anatomy of that particular region from available samples of multituberculates.

The palatal and nasal processes of the premaxilla of *Lambdopsalis* are similar in shape and proportion to those of *Djadochtatherium matthewi* Simpson, 1925. The nasal processes (or facial exposures) observed in these two taxa are proportionally more extensive than, and their palatal processes comparable to, those of described skulls of other multituberculates.

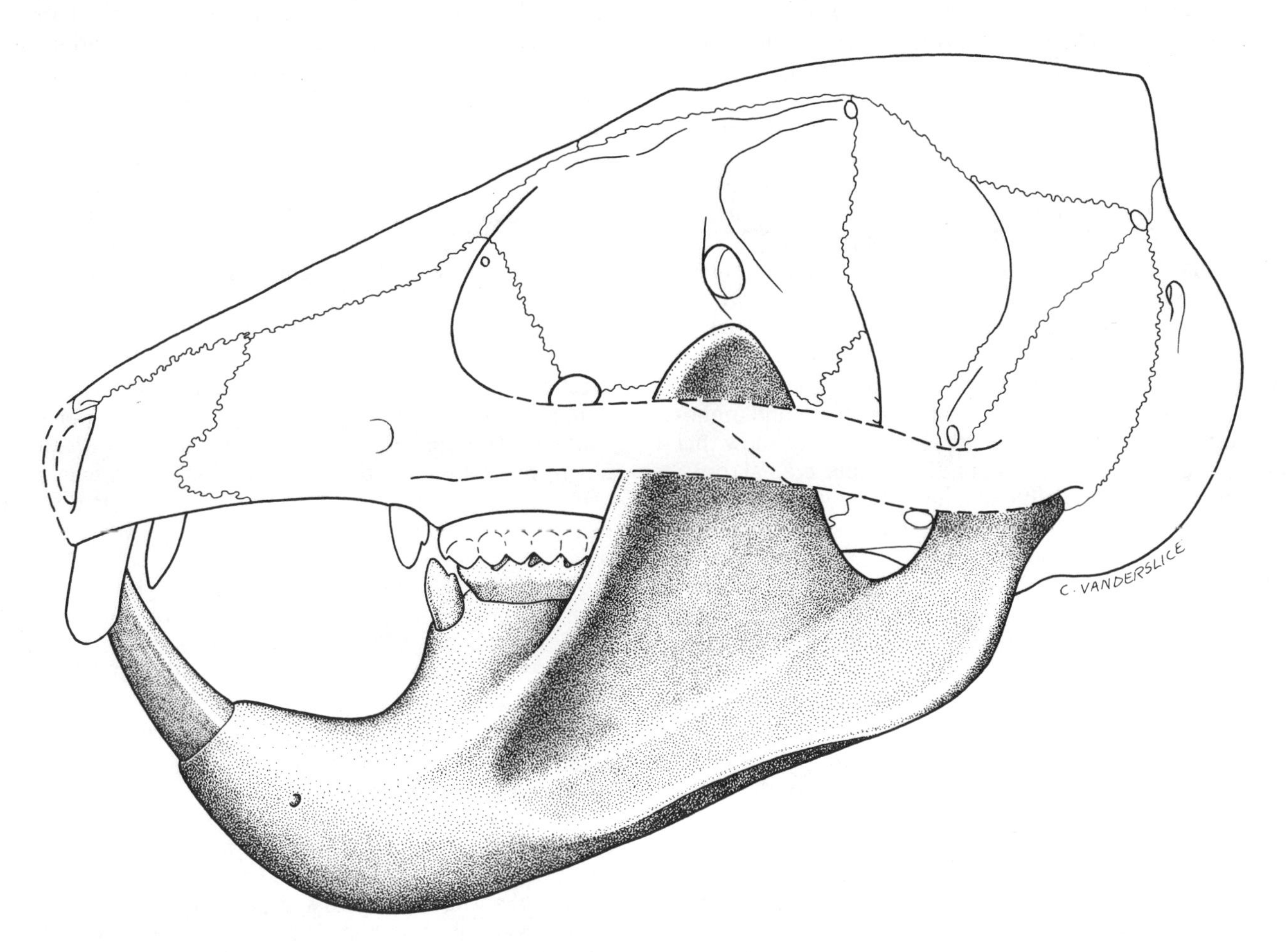

Figure 4. *Lambdopsalis bulla*, lateral view of composite reconstruction of skull.

Simpson (1925, 1937) did note the large, well-developed premaxilla in *Djadochtatherium*, and even regarded it as "decidedly aberrant" and of "peculiar development." However, he did not consider it of phylogenetic or functional significance. Even though the I2 of *Djadochtatherium* is broken off near its base, the diameter of its base indicates that, like I2 of *Lambdopsalis*, it is a much enlarged incisor. There seems to be a good correlation between a large, procumbent central incisor and an extensive premaxilla in mammals generally (*e. g., Dactylopsila* among marsupials, *Daubentonia* and *Plesiadapis* among primates, many rodents, and lagomorphs; see Gingerich, 1976). Related to the nipping bite, some carnivores (*e. g., Tomarctus*, borophagines, and hyaenas) also develop a large incisor and, in turn, an extensive premaxilla (S. Olson, 1985). This convergent evolution of large premaxillae in closely and distantly related groups therefore only implies possible functional rather than phylogenetic significance. To the extent that relative sizes and shapes of individual cranial bones are highly variable in response to changes in selective pressure, Erwin Stresemann metaphorically stated: "Bones are just like wax in the hands of evolution" (quoted in Lorenz, 1981, p. 95). As a result of the enlargement of the premaxilla, the frontal may come to contact the premaxilla. Similarly, this probably does not denote phylogenetic relatedness. However, Shoshani (1986) deemed premaxillo-frontal contact as a synapomorphy of "Glires." In fact, the premaxillo-frontal contact is characteristic of all the above-mentioned examples except for some multituberculates.

Like the sutures between other neighboring bony elements in skulls of *Lambdopsalis*, the premaxillo-maxilla suture is persistently visible through the life of the individual. The same is generally true of other multituberculates in which the structures are preserved. However, Kermack and Kielan-Jaworowska (1971) and Kielan-Jaworowska (1971) claimed that early fusion of cranial bones and early obliteration of sutures are a synapomorphy phylogenetically linking multituberculates and monotremes. Nevertheless, my personal observations show that sutures among neighboring bones of snouts, cranial roofs, and basicrania are visible in many Mongolian Late Cretaceous adult multituberculate skulls, and often are clearly shown with solid lines on the illustrations in descriptive papers. Therefore, in contrast to the monotreme condition, the occasional invisibility of sutures in some multituberculate skulls (especially in their more delicate regions, *e. g.*, orbitotemporal) may be due to the state of preservation rather than to early postnatal fusions.

The paired foramina on the palatal processes of premaxillae of *Lambdopsalis*, here designated as incisive foramina, also have been described in almost all available multituberculate cranial materials. Simpson (1925, 1937) prefered to call them anterior palatal foramina, but did consider them synonymous with incisive foramina. Kielan-Jaworowska (1971) and Kielan-Jaworowska and Dashzeveg (1978), however, dubbed them as palatine fissures (although these foramina make no contact with the palatine bones), and did not consider them to be "true" incisive foramina. It is not clear on what basis they rejected the openings as incisive foramina.

Nevertheless, all these authors seem to agree that these foramina were associated with Jacobson's organ (*i. e.*, vomeronasal organ, VNO). Though variable in size, shape, and exact position, incisive foramina occur in the premaxillary/maxillary region of the secondary palate of many mammalian groups, both fossil and extant, multituberculates included. Similar foramina are also present in therapsid skulls. Therefore, there seems to be little doubt about their homologous nature throughout the groups considered. Although recent detailed reviews on this subject have been consulted (Duvall, 1986; Duvall and others, 1983, and literature cited therein), the following points should be stressed:

1. In most groups of mammals, the vomeronasal organ persists as a functional chemosensory organ. It is basically a blind sac, buried toward the midline in the floor of the nasal cavity. In many rodents, it opens into the main nasal cavity. In most other cases, it literally separates from the nasal cavity in which it sits, and connects with the oral cavity via paired nasopalatine ducts, which pass through incisive foramina (Parsons, 1971; Romer and Parsons, 1977).

2. Vomeronasal organs disappear prior to completion of development in human and other higher primates, in many bats, and in various aquatic mammals such as cetaceans. Disappearance of VNOs in adult life, however, does not involve absence of incisive foramina.

With the above two premises, it may be concluded:

1. Presence of incisive foramina does not necessarily indicate the presence of VNO in the adult; therefore, we probably will never be sure if fossil forms such as multituberculates and morganucodontids indeed had VNOs in adult life, although they probably did.

2. As Duvall and others (1983, p. 29) pointed out, presence of an incisive foramen ". . . almost always indicates the presence of complete nasopalatine canals in mammals." But it does not always indicate ". . . an oral route to the paired VNOs." In those mammals which do not have VNOs as adults, the nasopalatine canals only transmit the terminal branches of the greater artery and nasopalatine nerve between the nasal cavity and the palate.

Despite the widespread trait of having incisive foramina in the multituberculate palate, there is a reported exception. Hahn (1969) reconstructed *Kuehneodon simpsoni* with paired incisive foramina, but later (1977) stated they are not visible because of strong lateral crushing of the skull. With new finds of *K. dryas* (which is also badly crushed), however, Hahn (*ibid.*) considered that incisive foramina are apparently missing.

Finally, it has been stated all too often in paleontological literature that the incisve foramen *housed* Jacobson's organ (*e. g.*, Kermack and others, 1981, p. 3). As discussed above, Jacobson's organ is housed in the floor of the nasal cavity. A foramen is a bidimensional orifice, not a three-dimensional space, and therefore really cannot house anything sac- or pouch-like.

"Septomaxilla"

Description

A septomaxilla can be discerned neither from facial exposure of the skull nor from the nasal cavity.

Discussion

No septomaxilla has been described in other multituberculate skulls, so its absence in *Lambdopsalis* is not surprising.

According to De Beer (1937, p. 442), "The septomaxilla (W. K. Parker, 1871), intranasal (Gaupp, 1904), or narial (Wegner, 1922), has been recognized in many reptiles and some mammals (Monotremes, Edentates) as a membrane-bone lying behind the external narial aperture and extending inwards as a plate overlying Jacobson's organ."

A septomaxilla also has been described in almost all therapsid groups, and usually has its facial exposure in front of the maxilla (Romer, 1956). It also has been reported in *Sinoconodon rigneyi* (see Patterson and Olson, 1961).

However, in spite of citations in secondary sources (*e. g.*, Romer, 1956 for *Oligokyphus;* Kemp, 1983 for *Morganucodon*), septomaxillary bones have been found neither in *Oligokyphus* nor in *Morganucodon*. The hypothesized presences of the bone were based: (1) in *Oligokyphus*, upon a contact surface on the maxilla (Kühne, 1956, p. 21); and (2) in *Morganucodon*, upon a facet on the premaxilla (Kermack and others, 1981). Both the contact surface and the facet were thought by the authors as hints for possible articulations with septomaxillary bones in those two taxa. However, the key element behind these speculations is that Kühne (1956, p. 22) reconstructed this bone in *Oligokyphus* based on its presence in *Tritylodon*, and Kermack and others (1981, p. 11), in turn, based on its presumed presence in *Oligokyphus* as well as its real presence in *Sinoconodon rigneyi*. Nevertheless, I am not suggesting that the indirect evidence for a septomaxilla in *Oligokyphus* and *Morganucodon* is not adequate to indicate their possible presence. Rather, I call attention to the fact that a septomaxilla per se has not been found so far in these two genera.

In visualizing a presumed facial exposure of septomaxilla in all advanced cynodonts and early mammals, Kemp (1983) suggested that exclusion of septomaxilla from the face is one of a few synapomorphies linking monotremes and modern therians. This view also was followed by McKenna (1987). It seems tenuous, however, when considering the following facts:

1. Septomaxilla is an obscure bony element, which may or may not occur in tetrapods of one group or another (Romer and Parsons, 1977). Its phylogenetic significance is poorly known.

2. Septomaxilla is excluded from the face, and thus is internal to the nostril in *Ctenosaura pectinata* among living reptiles (Oelrich, 1956).

3. In monotremes, a septomaxilla, if existed, only occurs in embryonic stages (Gregory, 1910; Jollie, 1962). De Beer and Fell (1936) reported that, even at the stage 5 (122 mm) of *Ornithorhynchus* embryo, the septomaxilla still ". . . is considerably larger than previously and is shaped like the alveolar process of the premaxilla, from which it is separated by the marginal cartilage and with which it will fuse when the marginal cartilage disappears" (p. 16). This is contrary to Kemp's (1983) interpretation that it is just ". . . reduced and internal to the nostril."

4. Although Fuchs (1911) and Jollie (1968) believed that the "septomaxilla" in monotreme embryos may be actually but a part of the premaxilla, they suggested that the septomaxilla of *Dasypus* is a real septomaxilla similar to that of living amphibians and reptiles. Jollie (1968) even commented that the identification of a septomaxilla in monotremes ". . . can also be traced to a desire to prove the primitive nature of the monotreme by the inclusion of this reptilain element" (p. 295). Considering retention of a septomaxilla in *Dasypus*, Jollie (1968, p. 296) further pointed out: "Disagreeable as it may be, one is forced to presume that a highly modified septomaxilla occurs in this single, highly specialized group of mammals." He also considered this odd appearance of the septomaxilla in *Dasypus* damaging to the comparative approach.

Recently, Shoshani (1986) also considered the absence of septomaxilla as a synapomorphy of Eutheria. However, until further information emerges, one should be cautious in using this variable, obscure morphological character for purpose of phylogenetic inference. A detailed study of the septomaxilla in mammals is being prepared by Wible, Hopson, and me.

Maxilla (Figs. 1, 3-5, 12, 17, 18)

Description

The maxillary bones are well preserved on the following specimens: IVPP V5429, V5429.10, V7151.2, V7151.5, V7151.8, V7151.50-55, V7151.79, and V7151.83.

The maxilla is an elongated, extensive bone forming the posterolateral part of the snout, a large part of the secondary palate, and basal part of the orbit. It articulates: anteriorly with the premaxilla; posteriorly with the alisphenoid; dorsally with the nasal, frontal, and orbitosphenoid; and ventromedially with the palatine and opposite maxilla. It bears three cheek teeth: one premolar (P4) and two molars (M1-2).

The maxillopremaxillary suture was described above under the heading of "Premaxilla." In dorsal view, the ascending process of the maxilla articulates with the posterior two-fifths of the lateral edge of the nasal, with a visible suture in most specimens. In lateral view, a prominent zygomatic process of the maxilla bounds the anterior rim of the orbit. Unfortunately, due to lack of a completely preserved zygomatic arch, in none of the specimens in the collection can the posterior limit of the process be pinpointed. In the bottom of the orbit a canal (here designated as the infraorbital canal) passes anteriorly through the root of the zygomatic process. The posterior entrance of the canal into the zygomatic process is at the anteroventral corner of the orbit (above the anterior tip of M1).

The anterior entrance of the infraorbital canal (*i. e.*, infraorbital foramen) is located at the anterior corner of the zygomatic arch (dorsal and anterior to P4). Only one infraorbital foramen on each side is observable in all the specimens except IVPP V7151.50, which has two infraorbital foramina on the right side. Of the two foramina, the more anterior and larger one is like that in all the other specimens; the more posterior and smaller one is separated from the former by a tiny bony bridge.

Within the orbit, the ascending process of the maxilla has an extensive dorsal contact with the frontal. The suture between the orbital process of the frontal and the ascending process of the maxilla forms a zigzaged line along the dorsal rim of a prominent groove (*i. e.*, sphenopalatine groove in the sense of Kielan-Jaworowska and others, 1986) at the bottom of the orbit. The groove leads forward into the infraorbital canal. Along the sphenopalatine groove, just posterior to the entrance of the infraorbital canal and dorsal to the midline of M1 in labial view, there is an elongated oval foramen, here designated sphenopalatine foramen. Dorsal to the anterior margin of the foramen, a suture ascends between the orbital process of the frontal and the zygomatic root of the maxilla.

The ascending process of the maxilla articulates posterodorsally with the orbitosphenoid at a position posterodorsal to the posterior margin of M2. Ventral to the posteroventral corner of the orbit and posterior to alveolar border of M2, the maxilla meets and is overlapped posteriorly and ventrally by the alisphenoid.

No lacrimal is discernible. However, a foramen is recognized at the anterodorsal corner of the orbit perforating the maxilla, at a position where a nasolacrimal foramen would be expected.

In palatal view, lateral to P4 and on the ventral surface of the zygomatic root of the maxilla, there is a pit of unknown function. The extensive palatal process of the maxilla constitutes the major part of the bony secondary palate. The transverse suture between the maxilla and the palatine is at about the level of the posterior one-third of M1. Along the suture (but basically on the maxilla) lies a foramen, usually referred to in other multituberculates as the major palatine foramen. The anteroposteriorly aligned suture between the maxilla and the palatine is more or less parallel to the tooth row, and runs posterolaterally to a sort of triple junction of the maxilla, palatine, and alisphenoid.

Within the nasal cavity and on the medial surface of the ascending process of the maxilla, anteroventrally-extended ridges and troughs are present, resembling those seen in modern mammalian skulls.

Discussion

As in other known multituberculate skulls, the maxilla of *Lambdopsalis* is an extensive dermal bone, forming most of the lateral and ventral sides of the rostrum and the anterior rim and floor of the orbit. An analogous condition is seen in many rodents. The implication of the maxillary contribution to the orbital mosaic will be discussed later under the heading of "Palatine."

Infraorbital foramen.—The maxilla of *Lambdopsalis* bears a single (with one exception already mentioned in description), average-sized infraorbital foramen at the anterior base of the zygomatic root. The foramen is similar to that seen in most known multituberculate cranial material except some paulchoffatiids and *Meniscoessus robustus*, in which two infraorbital foramina have been reported (Archibald, 1982; Clemens, 1973; Hahn, 1985).

The infraorbital foramen, often enlarged to a canal, occurs persistently in mammals, both living and fossil, transmitting the infraorbital artery and associated veins as well as the maxillary branch of trigeminal nerve forward from the orbit to the surface of the snout. It also has been reported in some cynodont therapsids where it is usually multiple (see Kemp, 1982). Until recently (Hahn, 1985; Kemp, 1983), however, little attention has been paid to the problem of its possible taxonomic value and/or phylogenetic implications in mammals except in rodents (see Hill, 1935).

Hahn (1985) examined the infraorbital foramen in multituberculates (especially paulchoffatiids), *Morganucodon* and *Priacodon ferox* among triconodonts, *Haldanodon exspectatus* among docodonts, and *Peramus tenuirostris* among eupantotheres. He regarded the tri- and bipartite structure of the infraorbital foramen as a symplesiomorphy of all pre-Cretaceous early mammals, and the undivided condition (*i. e.*, single foramen) as an apomorphy characteristic of younger multituberculates as well as of Metatheria/Eutheria. However, he assumed that ". . . its undivided condition has been evolved independently at least in Multituberculata and Metatheria/Eutheria" (p. 5). He did mention some disparity from this tidy morphocline in multituberculates (*e. g.*, the subdivided infraorbital foramen in the Late Cretaceous *Meniscoessus robustus*). It also should be noted that Hahn's survey of the character distribution among early mammals represents only a tiny fraction of once much more diverse groups (*i. e.*, one or two species out of each group). Even among the taxa mentioned, only one or two specimens of each taxon were available for study. Hence, the degree of individual variability cannot be ascertained for Hahn's study. Nevertheless, the variation seen in *Lambdopsalis* suggests that the individual variability of number of infraorbital foramina does exist.

Furthermore, Hahn (1985, p. 20) stated: "In younger multituberculates as well as in Metatheria/Eutheria only one foramen is present. . ." (referred to as the infraorbital foramen). But my survey of this character state among therians shows quite contrary results. For example, two infraorbital foramina are usually recognized in *Borhyaena* among marsupials, *Scelidotherium* among edentates, *Mammut americanus* among elephants, and *Dorcatherium* plus *Procamelus* among artidactyls. Three exist in *Diadiaphorus* among condylarths, and *Megalohyrax* among hyracoids. More than three are present in *Archaeolemur* among lemuroids and in many cetaceans (see also Gregory, 1920; Huber, 1934). It is also of interest to note that two infraorbital foramina are illustrated for

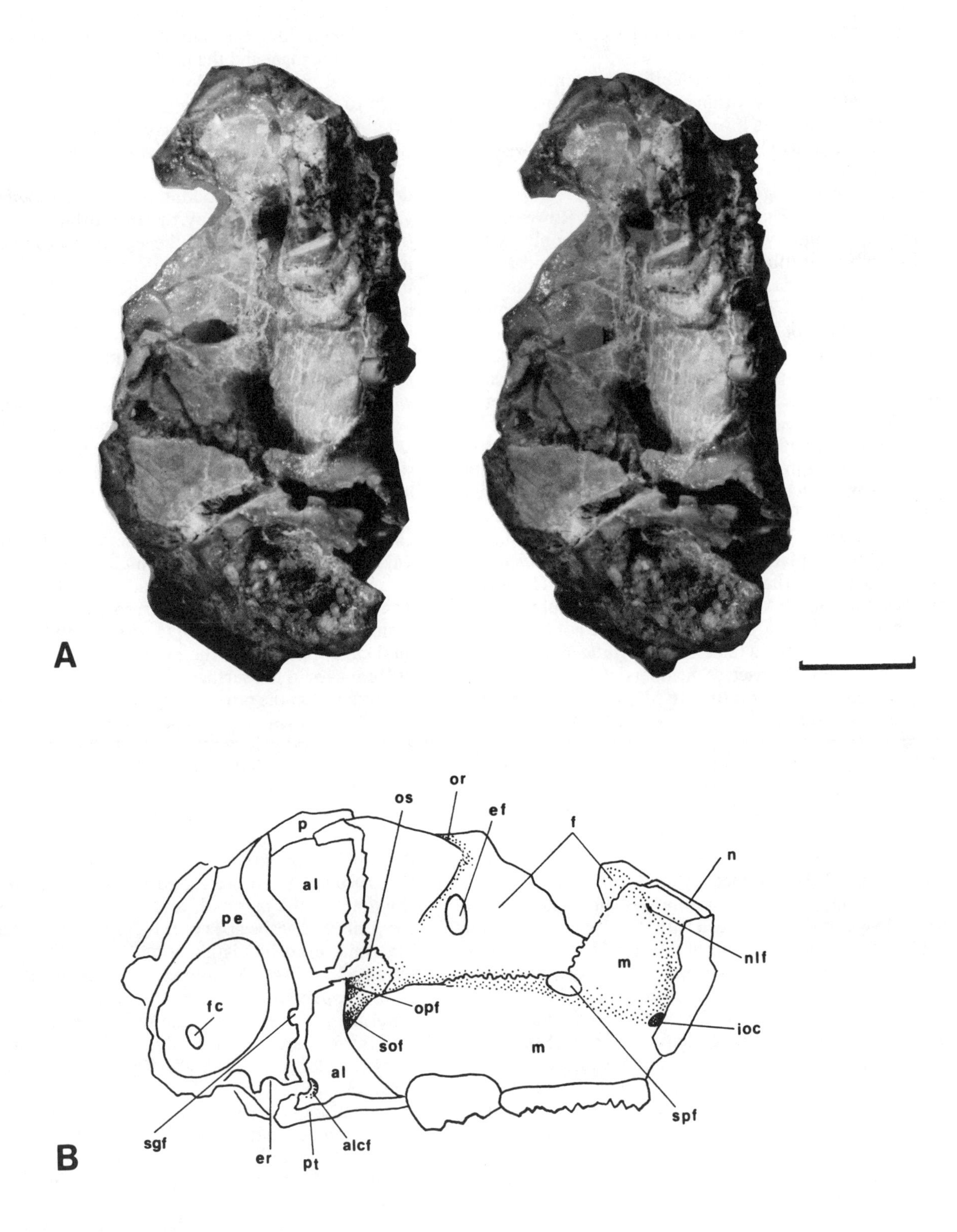

Figure 5. ***Lambdopsalis bulla*, V7151.81, lateral view of fragment of skull. *A*, stereophotograph, anterior to top; and *B*, explanatory drawing, anterior to right.**

an opossum (Jollie, 1962, Figure 3—5), but in all specimens of opossum available for my examination, there is only one. Major individual variation is indicated in addition to any regularity.

This high degree of variability in number of the infraorbital foramina within and among various groups of mammals renders a lower weight to its taxonomic and phylogenetic utility (Wheeler, 1986). Traditional wisdom has not focused much attention on this character in either taxonomy or phylogenetic reconstruction.

Based upon Hahn (1985), however, Rowe (1986) identified a single synapomorphy as shared only by the Guimarota multituberculates (*Paulchoffatia* and *Kuehneodon*) and Theria, a character suggested to be derived within Mammalia. This is ". . . a reduction in the number of the exits for the infraorbital canal onto the face" (Rowe, 1986, p. 224). Rowe conceived that the infraorbital canal in ancestral Mammaliamorpha (a rank created by Rowe to include all mammals in the traditional sense plus Tritylodontidae and Tritheledontidae), such as morganucodontids and monotremes, has three facial exits. Two are present in *Paulchoffatia* and *Kuehneodon*, and only one exists in most living therians. He further stated (1986, p. 224): "The shared presence of fewer than three exits for the infraorbital canal is here hypothesized as a synapomorphy placing *Paulchoffatia* and *Kuhneodon* [sic] in the lineage that includes living Theria." However, he was cautious in saying: "Because marsupials and placentals share many characters that have not been identified in the Guimarota taxa, *Paulchoffatia* and *Kuhneodon* [sic] are not themselves members of Theria, though they are closer to Theria than to Monotremata (see Fig. 1a)" (Rowe, 1986, p. 224-225). He also believed that this synapomorphy is the earliest character derived within Mammalia, and ". . . could therefore be taken as indicating the minimum age of Mammalia" (Rowe, 1986, p. 223), that is, Kimmeridgian. This presumed synapomorphy is a truly doubtful one, but I save this problem for later discussion.

Kemp (1983) examined the infraorbital foramen, or, more precisely, infraorbital canal system, in a different way. He was more concerned with the routes rather than the numbers of subdivisions of the infraorbital canal system. First, Kemp saw ". . . a basic similarity in the canal system distributing the maxillary branch of the trigeminal nerve to the snout in *all* cynodonts and mammals, with an infraorbital canal entering the floor of the orbit between the lacrymal bone above and the maxilla below" (p. 363, italics mine). This is only partly true. It is true that an infraorbital canal always leads forward from the floor of the orbit. However, it is not true that the canal always leaves the floor of the orbit between the lacrimal bone above and the maxilla below. In some mammals (*e. g., Ornithorhynchus*, most species of *Manis*, and *Lambdopsalis*), there is no lacrimal bone as such, and thus the infraorbital canal enters the maxilla alone. In some other mammals (*e. g., Tenrec*), the lacrimal bone has an extensive facial exposure, and is almost excluded from the orbit; the canal also enters the maxilla alone. Still in some other mammals (*e. g., Elephas indicus*), the lacrimal bone is relatively small and at the anterodorsal rim of the orbit; the canal enters the maxilla alone. In certain mammals (*e. g.*, lemuroids), the canal leaves the floor of the orbit between the jugal and the maxilla. Therefore, the infraorbital canal carrying blood vessels and nerves that lead forward to the surface of the snout along the floor of the orbit passes through whatever bone(s) it encounters, without deviating from its direct route.

Second, Kemp saw a difference between the cynodonts (including traversodontids) and *Oligokyphus*/ morganucodontids in the way the infraorbital canal passes through the maxilla. He believed: (1) that in cynodonts including traversodontids "the canal enters a large recess in the medial wall of the maxilla, the maxillary sinus, before sending off various branches to the external surface of the snout" (1983, p. 363); and (2) that in both *Oligokyphus* and morganucodontids, ". . . the posterior part of the infraorbital canal gives off three branches which go directly to the external surface of the skull" (*ibid.*, p. 363) without entering the maxillary sinus; and (3) the most posterior branch emerges at or near the suture between the lacrimal and the maxilla. Therefore, he considered differentiation of the infraorbital canal system in the latter style as a synapomorphy linking *Oligokyphus* and morganucodontids (thus ultimately linking tritylodontids and mammals).

Sues (1985) pointed out that no such distinction as Kemp claimed is borne out from fossil evidence. For example, in *Kayentatherium* (a tritylodontid), the infraorbital canal passes into the maxillary sinus before giving off branches to the surface of the snout, much as in other cynodonts. On the other hand, Sues argued that *Thrinaxodon* (". . . assumed to represent the stem from which all later cynodonts descended" [Kemp, 1982, p. 187]) already shows an independent "lacrimal" branch, similar to *Oligokyphus* and morganucodontids.

Furthermore, there are major differences in the canal system between *Oligokyphus* and *Morganucodon*, despite their shared similarity of having a "lacrimal branch." In *Oligokyphus*, the posterior part of the infraorbital canal is mainly in the lacrimal, and ". . . bent into a lateral convexity and at the apex of this curve a thin branch (*i. e.*, "lacrimal branch") is given off" (Kühne, 1956, p. 27, see also Text-Fig. 5A; note in these parentheses is mine). In *Morganucodon*, on the other hand, the infraorbital canal mainly enters the maxilla and leads more or less straight forward before sending off the lacrimal branch (*personal observation*, see also Kermack and others, 1981, Figs. 13 and 99A).

I am not arguing the significance of this difference. The difference may be purely due to different configurations of the skulls, and to the different proportions of the lacrimals between *Oligokyphus* and *Morganucodon*. Conversely, however, I do not see the significance of a similar possession of a lacrimal branch of the infraorbital canal in *Oligokyphus* and morganucodontids (among others). I agree with Sues that this feature is probably a symplesiomorphy. But I also offer a few comments: (1) I perceive methodological flaws in Kemp's analysis; (2) I stress the importance of comparative studies in character assessment; and (3) I stress the utility of character weighting in cladistic analysis.

Crucial importance of comparative studies in evaluating the reliability of a character.—In both above-discussed cases, attempts at using the infraorbital foramen or infraorbital canal system as a character to estimate relationship among the taxa concerned failed because of lack of knowledge of the character's distribution, both within the group and beyond the group. The more highly variable the character or the more randomly distributed the character, the more unreliable it is for phylogenetic inference, because enormous amounts of parallelism, or reversal, or retained plesiomorphy must be assumed. Hence, comparative methods remain powerful in recognizing homoplasy. In sum, the infraorbital foramen or canal is here regarded as an unreliable character, irrelevant in exploring the origin and relationships of mammals.

Undeniable utility of character weighting in cladistic analysis.—Despite some cladists' claim that character weighting is a confusing issue and should be avoided (Eldredge and Cracraft, 1980), even they must face the problem of character choice; an enormous number of characters can be identified from every specimen (*e. g.*, a single molar of a rodent). Indisputably, choice of the character itself involves a process of character weighting. Yet Kemp (1983, p. 381) explicitly stated: "In the formulation of the hypotheses which I present here, I have tried consciously to avoid preconceived ideas about which characters are more useful . . ." On the contrary, as soon as one tries to select a character, the worker cannot be free from subjectivity, no matter if he is pheneticist or cladist, or whether consciously or unconsciously doing so. As Wheeler (1986) pointed out, characters can be misinterpreted (and thereby errors can be introduced into the data set) by cladists as much as by taxonomists of other schools. Wheeler further stated: ". . . some recent cladistic analyses have resulted in levels of homoplasy of 80% or more" (p. 102-103). In my opinion (I will discuss this more fully later), Kemp's (1983) stimulating, though controversal, paper suffers from flaws in character assessment as I discussed before (*e. g.*, septomaxilla and infraorbital canal system). Sues (1985) made more extensive criticisms of Kemp's paper in this respect.

In fact, Kemp exclusively chose those characters presumed to be "only" shared by *Oligokyphus* and *Morganucodon* to aid in developing his argument of a sister-group relationship between tritylodontids and mammals. Many of these selected characters, however, are shared (in addition to *Oligokyphus* and *Morganucodon*) by other cynodonts, but not by other mammals. For example, considering the lacrimal branch of the infraorbital canal, Kermack and others (1981, p. 18) clearly stated: "We are unaware of the existence of this branch in any other mammals." Hence it is more likely to be a retained plesiomorphy in *Morganucodon*. However, without an adequate test (or weighting) of characters of this sort against their distributions among a wide range of representatives of the group concerned, Kemp made an inductive jump in assuming them as synapomorphies of tritylodontids and mammals. Procedural flaws thus crept in. As argued above, I believe that a thorough comparative study would have objectively weighed the characters, thereby precluding some of the errors. After all, a true synapomorphy is much more useful than a symplesiomorphic or homoplastic character in cladistic analysis. Should we really be expected to abandon character weighting altogether?

It may be argued that all characters in cladistic analysis are *a priori* nonconvergent characters (Eldredge and Cracraft, 1980). In practice, however, no warranty can be granted to this premise. Although cladism is a coherent system, the elegant methodology itself is not immune to the potential infections inherent in character choice. Moreover, blindly denying the necessity of character weighting, coupled with practically selecting only the "favorable" characters, leads to more confusion. Admitted or not, character weighting is as important to cladists as to any other school of systematics.

Lacrimal bone and nasolacrimal foramen.—There has been no report of presence of the lacrimal bone in Tertiary multituberculates, including *Lambdopsalis*. Among the Mongolian Late Cretaceous multituberculates, the lacrimal has not been recognized in *Djadochtatherium* (see Simpson, 1925) or the many taxa described by Kielan-Jaworowska (1971). Until claimed discovery of the lacrimal in *Nemegtbaatar* (see Kielan-Jaworowska, 1974), the reported absence of a lacrimal in multituberculates had in fact been used as a synapomorphy to argue for close relationship between multituberculates and monotremes (Kermack and Kielan-Jaworowska, 1971; Kielan-Jaworowska, 1971). At the other extreme, after the claimed discovery of a lacrimal in *Nemegtbaatar*, all other Mongolian Late Cretaceous multituberculates have been assumed to possess a lacrimal (Clemens and Kielan-Jaworowska, 1979; Kielan-Jaworowska, 1974; Kielan-Jaworowska and Dashzeveg, 1978; Kielan-Jaworowska and others, 1986). A lacrimal bone, however, was reported in the Jurassic *Paulchoffatia* (see Hahn, 1969). Considering presence of the lacrimal both in in-group and out-group (*e. g.*, cynodonts and morganucodontids) comparisons, absence of the lacrimal in *Lambdopsalis* as well as other multituberculates is here regarded as a secondary loss. I also consider that absence of lacrimal bone in monotremes is an independent secondary loss. Gregory (1920) recognized that the reduction or loss of the lacrimal in various groups of mammals may be due to differential development of its neighboring bony elements. As to the loss of the lacrimal in *Lambdopsalis* and a number of other multituberculates, it may be due to dorsal displacement by a well-developed zygomatic root of the maxilla.

Even though a lacrimal bone has been described at least in *Paulchoffatia* and *Nemegtbaatar*, a nasolacrimal (*i. e.*, lacrimal) foramen has been reported only in *Paulchoffatia* among multituberculates (Hahn, 1969). Similar to monotremes (see Jollie, 1962), *Lambdopsalis* has no lacrimal bone but, nevertheless, has a nasolacrimal foramen, which is presumed to have carried a lacrimal duct in life. This is inconsistent with Kühne's conclusion

(1956, p. 28) that: "So long as there is a lacrymal duct, the lacrymal can decrease but not disappear." In *Ornithorhynchus*, the nasolacrimal foramen opens between the frontal and the maxilla, but in *Lambdopsalis* it perforates the maxilla alone.

Sphenopalatine foramen.—The sphenopalatine (or internal orbital) foramen occurs in mammals in general, including multituberculates. It is usually located in the rostroventral part of the orbit. Due to complexity of the orbital mosaic in different mammalian lineages, the sphenopalatine foramen may be bordered by one or several bones of the anterior orbital mosaic. As examples, among rodents it is within the maxilla in *Eutypomys* and in many geomyids (Wahlert, 1977, 1978), and bounded by the frontal, maxilla, and palatine in florentiamyids (Wahlert, 1983). Among multituberculates, the sphenopalatine foramen is bounded by: maxilla and the orbitosphenoid in *Ptilodus* (see Simpson, 1937); the junction of the maxilla, frontal, orbitosphenoid, and palatine in *Nemegtbaatar* (see Kielan-Jaworowska and others, 1986); the maxilla and palatine in *Ectypodus* (see Sloan, 1979); and in *Lambdopsalis* the frontal and maxilla within the sphenopalatine groove. As mentioned above, this high diversity of position among different taxa is correlated with high complexity of the orbital mosaic. Therefore, a superficial similarity of the sphenopalatine foramen's topographic relationship with the surrounding bones in different taxa does not necessarily denote phylogenetic relatedness. It may often be a homoplastic character. The principal functions of the sphenopalatine foramen in living mammals are to transmit the nasal and palatal nervous branches from the sphenopalatine ganglion and the sphenopalatine artery to the interior of the nasal cavity (to innervate and supply mucous membranes of the nose and palate; see also Kielan-Jaworowska and others, 1986; Story, 1951; Wang and Lung, 1984). I see no reason to doubt similar functions in multituberculates.

Major palatine foramen.—As in *Lambdopsalis*, a small foramen is present on or near the transverse suture between the palatal process of the maxilla and the palatine in almost all cynodonts, *Morganucodon, Ornithorhynchus, Kamptobaatar*, and many modern therians. It has been given different names by various authors. For example, it is called the greater palatine foramen in *Morganucodon* (see Kermack and others, 1981), anterior palatine foramen in *Ornithorhynchus* (see Jollie, 1962) and *Arctomys* (see Yapp, 1965), the posterior palatine foramen in many rodents (Wahlert, 1977, 1978, 1983, 1985), and major palatine foramen in *Kamptobaatar* (see Kielan-Jaworowska, 1971; Kielan-Jaworowska and others, 1986). It is called the rostral end of the palatine canal in *Felis domesticus* by Walker (1986). Kühne (1956) simply called it the anterior opening for the palatine nerve. I recommend using the term "major palatine foramen" for the following reasons: (1) the greater palatine foramen is a term basically used in human anatomy; (2) the anterior and the posterior foramina are equally confusing when some authors choose to call the incisive foramen as the anterior palatine foramen and some do not; (3) the major palatine foramen has been established in literature at least for multituberculates; and (4) it is as good as any other proposed term and yet avoids confusion.

The relatively constant position and presumably similar function of the foramen have not been masked by the terminological confusion. Based upon its function in modern mammals, it has been interpreted in fossil forms as the passage for the greater palatine nerve and the descending palatine artery, both to emerge on the surface of the palate.

Although considerable attention has been paid to the major palatine foramen, relatively little has been written about the entry of the greater palatine nerve from the orbit. In many eutherians (*e. g.*, rodents, carnivores, *etc.*), a small foramen is present near, but separate from, the sphenopalatine foramen. The small foramen is sometimes called the posterior end of the palatine canal. It is through this small foramen that the greater palatine nerve leaves the orbit. When this foramen is present, usually there is a palatine canal within the palatine or between the palatine and the maxilla, carrying the greater palatine nerve, which is totally isolated from the nasal cavity. In such a case (*e. g., Felis domesticus*), the greater palatine nerve branches off the sphenopalatine nerve before the sphenopalatine nerve reaches the sphenopalatine foramen to enter the sphenopalatine ganglion (Reighard and Jennings, 1935). When the small, separate foramen is absent, the greater palatine nerve leaves the orbit via the sphenopalatine foramen. In the latter case, the greater palatine nerve branches off the sphenopalatine nerve after the sphenopalatine nerve traverses the sphenopalatine ganglion. In other words, the greater palatine nerve may be considered as one branch from the ganglion. In such a case, there is often a groove on the medial surface of the ascending process of the palatine instead of a completely separate bony palatine canal. Hence, the greater palatine nerve is partly exposed to the nasal cavity after entering the sphenopalatine foramen. A similar condition has been described in *Oligokyphus* (see Kühne, 1956, p. 56-57), and *Morganucodon* (see Kermack and others, 1981, p. 53-58). This is also the case with *Lambdopsalis*, except for the greater palatine groove being on the maxilla due to exclusion of the palatine from the anterior part of the orbit.

A similar arrangement can be observed in many marsupials and insectivores. The only difference in these cases is that, instead of a small major palatine foramen, a rather large palatal vacuity allows the greater palatine nerve passage onto the surface of the palate. This interpretation may be extended to the other multituberculates having a large palatal vacuity but without a separated foramen for exit of the greater palatine nerve to the palate. In fact, Simpson (1937, p. 756) postulated that a major palatine foramen (*i. e.*, his "palatine foramen") ". . . is probably an opening within the posterolateral part of the rim of the vacuity."

While Kermack and others (1981) admirably described the foramina and reasonably explained them as being passages for certain nerves, their functional interpretations of the nervous components contained fundamental misconceptions. For example, they stated (p. 55): "Although in living mammals the greater and lesser palatine nerves appear to arise from the sphenopalatine ganglion, they have no functional connection with it. Both greater and lesser nerves are branches of the maxillary ramus of the trigeminal nerve (V) and are mainly sensory in function. The sphenopalatine ganglion is part of parasymphathetic [sic] system."

As far as known in living mammals, all lesser palatine nerves and a majority of greater palatine nerves are connected with the sphenopalatine ganglia. Although these two nerves consist of a large number of sensory fibers from the sphenopalatine nerve (a branch of V2), they also receive contributions of postganglionic sympathetic nerve fibers from the facial nerve (VII), and of postganglionic parasympathetic fibers from the cells of the sphenopalatine ganglion (Romanes, 1981). In other words, after the branch of the sphenopalatine nerve traverses the ganglion, its sensory nerve fibers run in the same fibrous sheath as the fibers of the autonomic nervous system, and are distributed to the same territories. Hence, the greater and lesser palatine nerves are no longer the "pure" branches of the maxillary ramus of the trigeminal, and instead contain the fibers from peripheral (*e. g.*, cranial) and autonomic (*i. e.*, sympathetic and parasympathetic) systems. They are, therefore, sensory, secretomotor, and vasomotor to the territories they supply. Accordingly, even though the sensory fibers "inherited" from the sphenopalatine nerve are not functionally related to cells of the sphenopalatine ganglion, the greater and lesser palatine nerves have rather close functional connections with the ganglion.

To sum up, based upon both in-group and out-group comparisons, the confluence in *Lambdopsalis* of the foramen for entry of the greater palatine nerve into the palatine canal (or groove) with the sphenopalatine foramen, along with presence of the greater palatine groove instead of the palatine canal, seems to be a plesiomorphy for mammals in general.

Nasal (Figs. 1, 2, 4, 6, 12, 17)

Description

The nasal can be seen well in the following specimens: IVPP V5429.2, V5429.10-12, V7151.50, V7151.55, V7151.79, V7151.82, V7151.83, and V7151.85.

The nasal is a large bone, narrow anteriorly and gradually expanded posteriorly. But at about the posterior one fifth, it tapers posteriorly. Although the anterior tip of the nasal is broken off in all of the above-listed specimens, preserved parts (especially seen in V7151.79) indicate that it almost reaches the anterior limit of the snout. The paired nasals form the roof and the most dorsal rim of the lateral walls of the nasal cavity. They are in sutural contact with each other medially, with the premaxillae both medially and laterally, with the maxillae laterally, and with the frontals laterally and posteriorly. The tip of the posterior process of the nasal contacts the tip of the anterior process of the parietal. The posterior process of the nasal reaches the anterior orbital rim. All the sutures are visible throughout life. The suture between the nasal and frontal is splintery, and the nasal is overlapped by the frontal.

There are seven to eight foramina (here called "nasal foramina," after Simpson 1937) in each nasal bone. Four larger ones (*a, b, c, d*) are anteroposteriorly aligned, and the most posterior one (*d*) is the largest as an elongated oval. Among three smaller foramina (*e, f, g*), one (*e*) is medial to and between the first two anterior large foramina (*a* and *b*), another (*f*) medial to the penultimate (*c*), the other (*g*) either medial or lateral to the most posterior, largest one (*d*). These foramina are arranged symmetrically to those in the opposite nasal.

Although the ventral surface of the nasal is not exposed in any of the above-listed specimens, five major anteroposteriorly-extended grooves on the endocast of the nasal cavity (V5429.13) indicate presence of five correspondent ridges on the ventral surface of the nasals. These ridges apparently are continuous with those on the ventral surface of the frontals. The middle ridge is along the suture zone of the nasals, here designated as the median ridge of the nasal.

Discussion

Peculiar development of the nasal.—As in most other known multituberculate skulls, the nasals in *Lambdopsalis* are also proportionally large compared with those of many other mammals. Simpson (1937) commented that "the peculiar development of the premaxillae and of the nasal" are "decidedly aberrant" (p. 732). The shape of the nasal in *Lambdopsalis* resembles *Djadochtatherium* more than other multituberculates from which the nasal is known.

The nasal bone apparently became greatly enlarged in advanced cynodonts as well as in early mammals, correlated with enlargement of the nasal cavity. Moore (1981, p. 241) speculated: "This enlargement was undoubtedly related to the increased respiratory requirements associated with the emergence of a mammalian type of metabolism. It must also have been of advantage in allowing an increase in the size of the peripheral olfactory apparatus, a factor which may well have been crucial in the emergence of the earliest mammals which, it is believed, were nocturnal creatures and greatly dependent, therefore, upon the possession of an acute sense of smell." This seems to be a plausible explanation for Simpson's observation about "the peculiar development" of large nasals in multituberculates.

In addition, the nasal is overlapped by the frontal in *Lambdopsalis*. This is reminiscent of *Ursus* (see Gregory, 1910, p. 424), but contrasts with the usual condition seen in other multituberculates and most mammals, in which the frontals are overlapped by the nasals.

Nasal foramina.—Except *Paulchoffatia*, all reported multituberculate cranial materials in which the nasal bone is preserved show presence of what Simpson (1937) called "vascular foramina" or "nasal foramina." Although these foramina differ among species in number (1 to 7 or 8), shape (round to elongated), and arrangement (symmetrical *versus* asymmetrical), their consistent presence

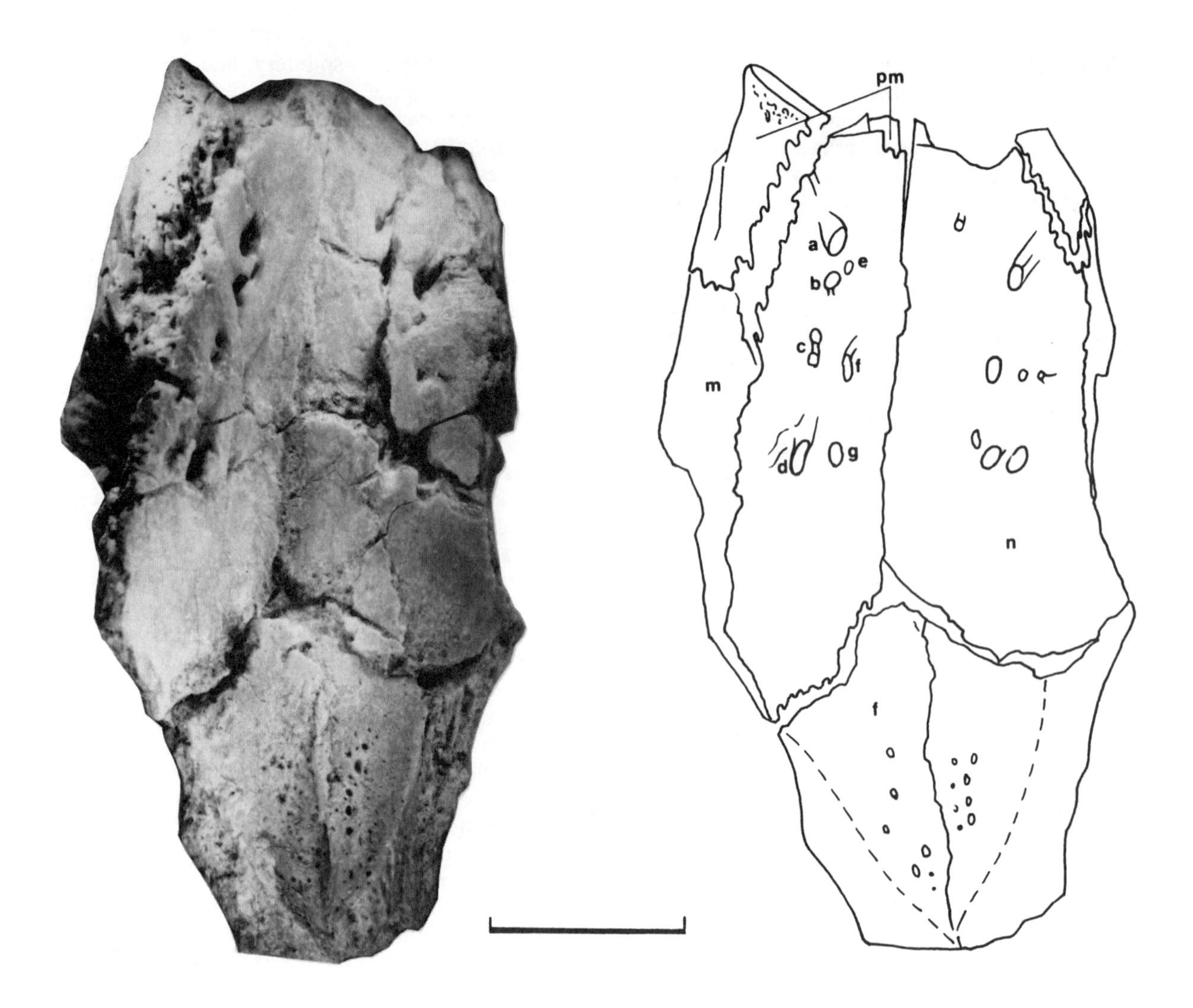

Figure 6. ***Lambdopsalis bulla*****, V7151.82, dorsal view of fragment of skull (anterior to top); nasal foramina coded by *a, b, c, d, e, f,* and *g* on nasal.**

in multituberculates led Kielan-Jaworowska (1971) to think that they are characteristic of the Multituberculata as a whole. Failure to identify such foramina in *Paulchoffatia* may be due to poor preservation of the specimens. Nasal foramina were not described in the specimen of *Taeniolabis* by Broom (1914) due to poor preservation, but they are present in a better preserved skull of *Taeniolabis* being studied by Simmons (*personal communication*). Based upon similar foramina on the nasal bones in recent lizards (Oelrich, 1956), Kielan-Jaworowska (1971; Kielan-Jaworowska and others, 1986) suggested that the foramina served similar functions in multituberculates, in transmitting branches of the lateral ethmoid nerve. I can neither confirm nor refute this extrapolation in functional interpretation. However, I do recommend use of "nasal foramina" instead of "vascular foramina" (both being Simpson's terminology) for purposes of description in multituberculates; the latter term denotes a speculated function but does not indicate topographic relation with the nasal bone. It also should be noted that somewhat similar foramina are present on the nasals of some therapsids (*e. g., Procynosuchus*, see Kemp, 1979) and *Morganucodon* (see Kermack and others, 1981). But the foramina in therapsids and *Morganucodon* appear to be less regularly distributed, and differ from those seen in multituberculates in arrangement (note the complex yet regular distribution of these foramina in *Lambdopsalis*). As Kielan-Jaworowska (1971, p. 25) noted: "such foramina are quite unusual for mammals." Further study may suggest them as synapomorphy of multituberculates.

For discussion of features on the ventral surface of the nasal, see section "Nasal Cavity" below.

Frontal (Figs. 1, 2, 6-9, 11, 12, 17)

Description

The frontal is well preserved in the following specimens, upon which the description is based: IVPP V5429.3, V5429.4, V5429.14, V7151.5, V7151.50-52, V7151.55, V7151.79, and V7151.82-86.

The frontal may be divided into two parts, the fron-

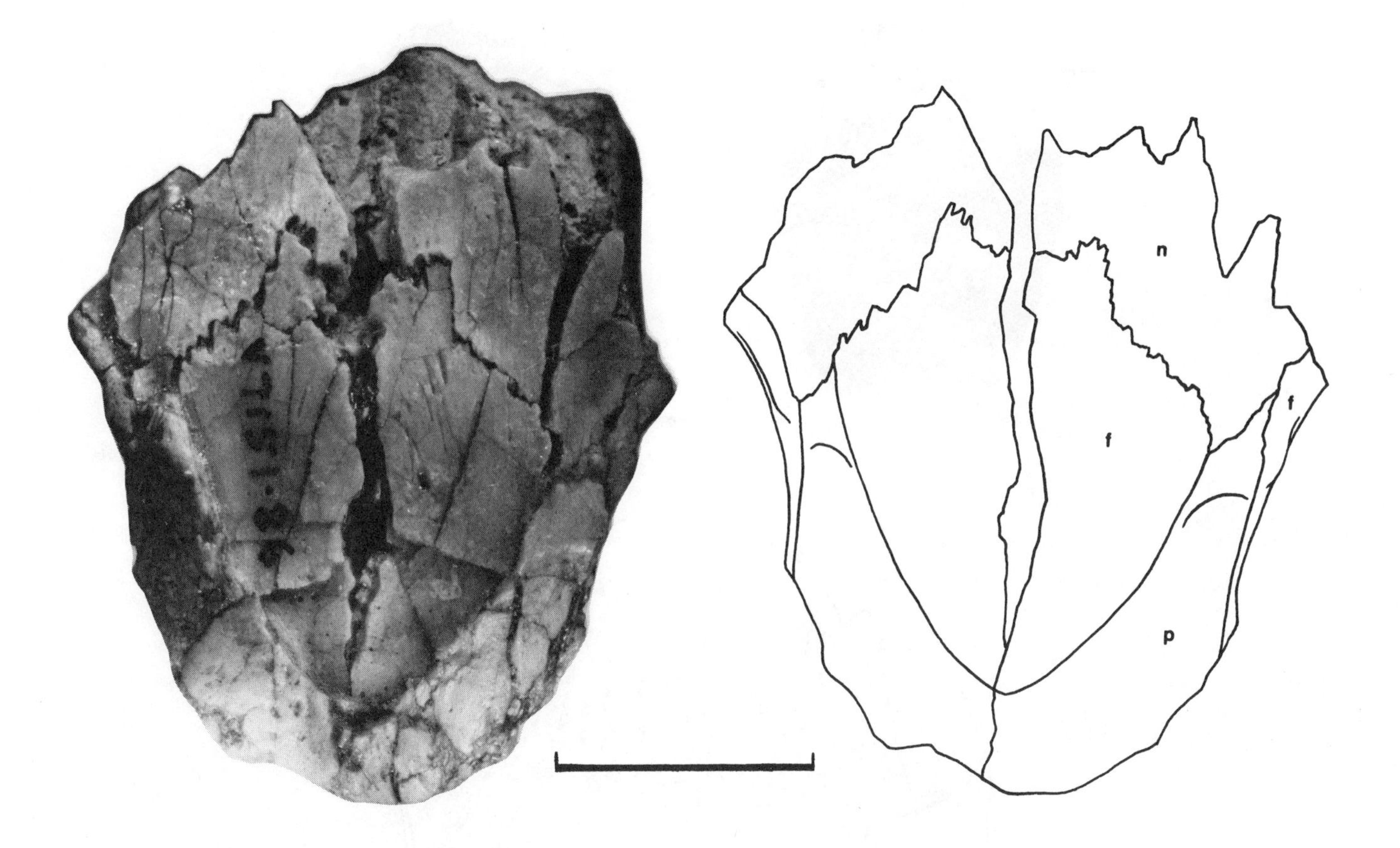

Figure 7. ***Lambdopsalis bulla*, V7151.86, dorsal view of fragment of skull. Anterior to top.**

tal plate (more or less horizontal) and orbital process (more or less vertical).

The frontal plate forms the posterior part of the roof of the nasal cavity and the anterior part of the roof of the cranial cavity. It overlaps the nasal anteriorly and is overlapped by the parietal both laterally and posteriorly. The exposed part of each frontal plate is roughly triangular, with an anterior acute process projecting between the nasals and a posterior acute process between the parietals. The dorsal surface of the frontal plate is almost flat, with numerous tiny pits and anteromedially-extended shallow furrows. The frontal plates meet each other medially in a straight mid-dorsal suture. The lateral border of the frontal plate is thickened into a ridge (similar to that frequently seen in other mammals), and here designated as the supraorbital crest. At the posterior part of the supraorbital crest, a short but distinct postorbital process projects laterally at a position above the anterior margin of M2 in lateral view. The greatest constriction of the skull is posterior to the postorbital process, at a level with the end of the posterior acute process of the frontal plate.

The orbital process of the frontal is the orbital exposure of the frontal, descending from the supraorbital crest into the orbit. The orbital process is laterally concave, and forms the major part of the medial wall of the orbital fossa. It is in sutural contact with the ascending process of the maxilla anteriorly and ventrally, and with the alisphenoid and orbitosphenoid posteriorly.

At the anterodorsal corner of the orbit, a small, elongated oval fossa is present on the orbital process of the frontal, just ventral to the supraorbital crest. I term the fossa the "orbitonasal fossa." Anterior to the suture between the orbital process of the frontal and the alisphenoid, there is a ridge (here designated as the "orbital ridge") on the orbital process, descending from the supraorbital crest to the triple junction of the orbital process of the frontal, alisphenoid, and orbitosphenoid. The orbital ridge is slightly convex anteriorly. Anterior to the orbital ridge, there is a large foramen (here designated as ethmoid foramen) which is bounded posteriorly by the middle part of the ridge. Immediately dorsal to the ethmoid foramen is a posterodorsally-extended groove (here designated as the "orbital groove"), which ends at a foramen (termed postorbital foramen) at the triple-junction of the frontal, alisphenoid, and parietal.

Features of the ventral surface of the frontal plate of the frontal are similar to those seen in *Morganucodon*, superbly described by Kermack and others (1981, p. 27-29). A laterally convex ridge arises at the inner surface of an orbital constriction anterior to the postorbital process, and converges anteromedially toward the suture between the two frontal plates. This convergence results in an anteriorly convex transverse ridge, which forms part

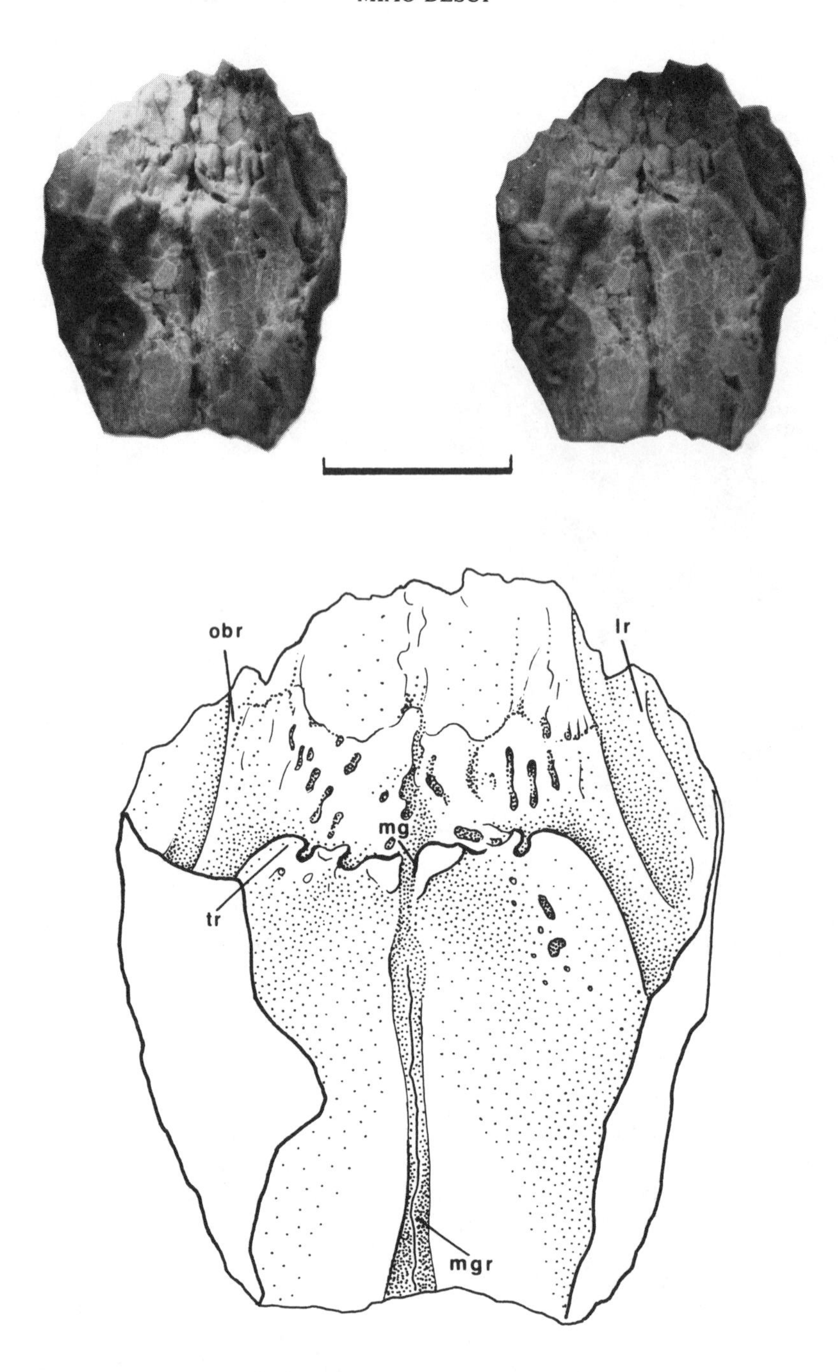

Figure 8. ***Lambdopsalis bulla*****, V7151.84, stereophotograph, ventral view of fragment of skull showing internal surfaces of parts of frontal and parietal. Anterior to top.**

of the boundary between the cranial and nasal cavities. The transverse ridge, however, is discontinuous, interrupted by five gutters at the anterior convexity of the ridge. The median gutter is the anterior termination of a deep median groove on the frontal roof of the cranial cavity. Anterior to the median gutter and continuous with the median groove, there gradually arises a median ridge, which is the posterior continuation of the median ridge of the nasal.

At the posterolateral corner of the transverse ridge, a deep gutter separates the transverse ridge from an oblique ridge. The oblique ridge runs more or less parallel to the median ridge, but with a slight anteromedial deviation. Lateral and parallel to the oblique ridge is the lateral ridge. Both the oblique and lateral ridges have their anterior continuations with the correspondent ridges of the nasal. Between the median and oblique ridges, there is a wide, shallow trough. A narrower and deeper trough is between the oblique and lateral ridges. These ridges and troughs are within the nasal cavity, terminating at the transverse ridge. Both along the transverse ridge and anterior and posterior to the ridge, there is a narrow zone

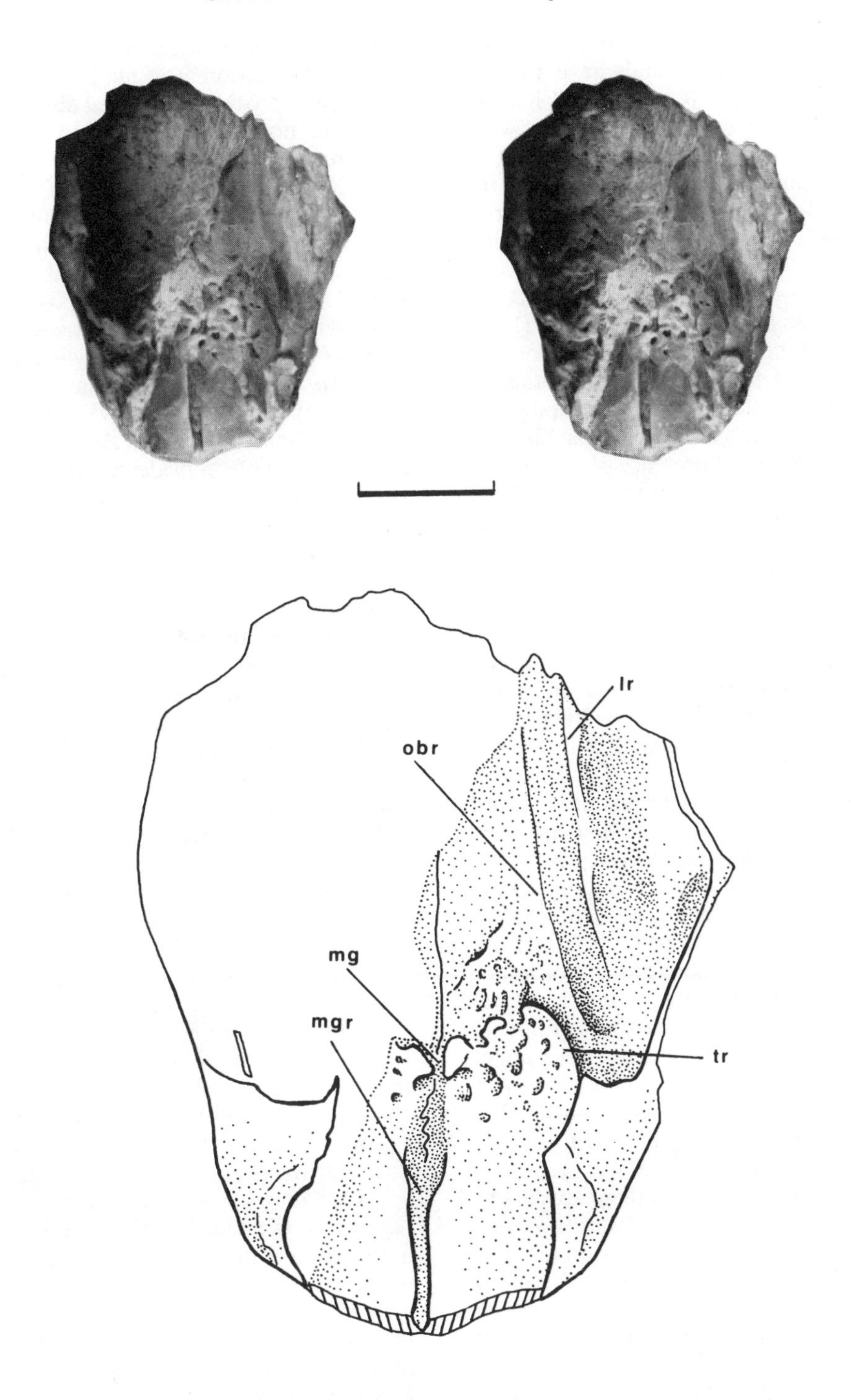

Figure 9. ***Lambdopsalis bulla*, V7151.86, stereophotograph, ventral view of fragment of skull showing internal surfaces of parts of nasal, frontal, and parietal. Anterior to top.**

of rough surface with numerous irregularly distributed pits. Immediately posterior to the rough surface, the frontal becomes greatly thickened. The broken sections at this level seen in IVPP V7151.51 and V7151.86 reveal well-developed frontal sinuses. A small opening anterior to the sinus on each side communicates with the nasal cavity.

The inner surface of the orbital process of the frontal is marked by several anteroventrally-running ridges.

Discussion

In the usual mammalian condition, the frontal is a dermal bone covering the cranial roof in the orbital region. The shape of the frontal plate of *Lambdopsalis* is more similar to that of *Djadochtatherium* than to other described multituberculates. The frontal plates taper posteriorly and insert between the parietals approximately as in the North American multituberculates *Taeniolabis* and *Ptilodus*. However, this is different from all the other known Asian taxa, in which the frontals have a transversely-aligned posterior margin (Clemens and Kielan-Jaworowska, 1979).

The extensive orbital exposure of the frontal in *Lambdopsalis* is probably related to the degree of development of other neighboring elements in the orbital mosaic, which will be discussed in the section "Palatine."

Postorbital process.—*Lambdopsalis* has a short but distinct postorbital process of the frontal. Simpson (1937) did not describe, but did indicate in the illustrations, the presence of a postorbital process of the frontals of *Taeniolabis* and *Ptilodus*. Sloan (1979) also indicated presence of a postorbital process in *Ectypodus* in his illustrations. Therefore, it seems that presence of a postorbital process of the frontal is characteristic of all post-Mesozoic multituberculates in which cranial materials are known. However, a postorbital process of the frontal may be absent in Mongolian Late Cretaceous multituberculates. Even though Kielan-Jaworowska (1971) identified the posterior termination of the supraorbital crest as a postorbital process in *Kamptobaatar* and *Sloanbaatar*, it probably is not the postorbital process in a conventional sense. Ordinarily, the postorbital process in mammals is a process of the frontal. But in Kielan-Jaworowska's reconstruction, it is a process of the parietal. In addition, it is far posterior to the postorbital constriction, which always is posterior, rather than anterior, to the postorbital process in mammals. Consequently, she might have interpreted the eyes of the Mongolian Late Cretaceous multituberculates as having been larger than they really were. Poor preservation of the Kimmeridgian multituberculates renders impossible identification of the postorbital process. Thus the polarity of this character within multituberculates is difficult to determine.

Among Mesozoic mammals in which cranial materials have been described, Kermack and others (1981) suggested the frontal in *Morganucodon* has no postorbital process. Concerning *Sinoconodon*, they pointed out: "Although Patterson & Olson (1961:143, fig. 3) show *S. rigneyi* with a short postorbital process there appears to be no positive evidence for this. As *Morganucodon* definitely has no postorbital process it is reasonably certain that the same holds true for *Sinoconodon*" (Kermack and others, 1981, p. 27). While a logical flaw exists in Kermack and others' argument, personal observations confirm that Patterson and Olson's reconstruction of a postorbital process in *Sinoconodon* is correct. Admittedly, the materials available to Patterson and Olson were difficult to deal with due to poor preservation. The same is true of the cranial materials of *Morganucodon*. The part of the frontal (*M. oehleri*) in C.U.P. 2320 at which a postorbital process would be expected is broken off. Remaining materials of the frontal of *Morganucodon* consist of isolated fragments. It remains uncertain whether *Morganucodon* indeed lacks a postorbital process. It also should be noted that Crompton and Sun (1985, Fig. 1) reconstructed the skull of *Sinoconodon changchiawaensis* with a postorbital process in dashed lines. But it is not clear upon what basis this aspect of the reconstruction was made. In any case, even if a postorbital process was indeed absent in *Morganucodon*, it would not necessarilly mean it was also absent in *Sinoconodon*.

To turn the argument around, it seems probable that the postorbital process of the frontal is a neomorph in mammals. It would not be surprising if the earliest mammals such as *Morganucodon* did not have it. If this is indeed the case, absence of the postorbital process of the frontal in Mongolian Late Cretaceous multituberculates may be a symplesiomorphy, whereas presence of it in *Lambdopsalis* and other Tertiary multituberculates may be a synapomorphy. However, this is only an assumption, pending further test. As an aside here, I note that Novacek (1980) considered absence of a postorbital process as primitive in his phylogenetic analysis of eutherians.

Orbitonasal fossa and ethmoid foramen.—an orbitonasal fossa was first noted and described by Kielan-Jaworowska (1971) for *Kamptobaatar* and *Sloanbaatar*. Subsequently, an orbitonasal fossa also was described in *Chulsanbaatar* and *Nemegtbaatar* (see Kielan-Jaworowska and others, 1986). It may vary in number (from 1 in *Kamptobaatar* and the other taxa mentioned to 3 in *Sloanbaatar*) and shape (from round in most taxa mentioned here to elongated-oval in *Lambdopsalis*). Kielan-Jaworowska (1971) noted it as a peculiar feature, unknown in other mammals. Kielan-Jaworowska (1971) speculated that its function was to house a Harderian gland. Kielan-Jaworowska and others (1986) thought it was for accommodation of an unknown gland.

Kielan-Jaworowska (1971) also noted a minute foramen in the posterior part of the orbitonasal fossa in *Kamptobaatar*. She designated it as the orbitonasal foramen, following Jollie's (1962) usage for monotremes. However, this minute foramen has not been found in other Asian multituberculates, including *Lambdopsalis*. Also, personal observation of the unnatural rim of the foramen leads me to suspect that the foramen is a preparation artifact.

Due to generally poor preservation of the orbital region, an ethmoid foramen can be recognized neither in the Kimmeridgian multituberculates from Portugal nor in the North American Paleogene multituberculates. It was, however, identified in Mongolian Late Cretaceous taxa (Kielan-Jaworowska, 1971; Kielan-Jaworowska and others, 1986). Differences in the ethmoid foramen between *Lambdopsalis* and the Mongolian Late Cretaceous taxa lie in position and size. The ethmoid foramen of *Lambdopsalis* is proportionally larger, piercing the orbital process of the frontal anterior to the orbital ridge. In contrast, the ethmoid foramen of the Mongolian Late Cretaceous multituberculates is smaller, passing through the orbitosphenoid posterior to the orbital ridge. Positions of the ethmoid foramen in modern mammals vary within families from virtually any location on the orbital process of the frontal to anywhere dorsal to the optic foramen on the orbitosphenoid. In some cases, the ethmoid foramen is situated on the suture zone between these two bones. Therefore, it follows that even if the difference in position of the ethmoid foramen between *Lambdopsalis* and other Mongolian taeniolabidiids were real, this character alone would not exclude their probable, close affinity at a familial level. However, it is uncertain if the ethmoid foramen in the Mongolian taxa emerges through the orbitosphenoid. Additionally, the orbitosphenoid of the Mongolian Late Cretaceous multituberculates, as reconstructed, seems to be unusually large for mammals. Hence I suspect it might have been exaggerated in reconstruction.

The ethmoid foramen of *Lambdopsalis*, however, is indeed unusually large. A similar condition is known only in *Morganucodon* among early mammals (Kermack and others, 1981). Kermack and others attributed the unexpected size of the ethmoid foramen in *Morganucodon* to its location in an area of unossified braincase wall in life. Such is certainly not the case in *Lambdopsalis*; its ethmoid foramen has a well-defined, round margin.

The ethmoid foramen in modern mammals transmits the ethmoidal branch (a continuation of the nasociliary branch) of the ophthalmic nerve (V1) plus the ethmoid artery and vein. Similar functions would be expected for the ethmoid foramen in *Lambdopsalis*. The large size of the foramen in *Lambdopsalis* may be related to the radii of the ethmoid nerve and blood vessels.

Frontal sinus.—The frontal sinus is a system of pneumatic chambers developed within the frontal bone in various groups of mammals, both living and extinct. The sinus usually communicates with the nasal cavity. The extensive studies of the paranasal sinuses (a term to include frontal, maxillary, and sphenoidal sinuses) by Paulli (1900*a,b,c*) remain as standard references. However, a more recent review of this subject is by Moore (1981).

According to Moore (1981), Paulli (1900*c*) found no evidence of presence of paranasal sinuses in monotremes and many marsupials. Thus it has been assumed that the presence of the paranasal sinuses may be an eutherian synapomorphy. However, the reported presence of maxillary sinuses in some gorgonopsids (Kemp, 1969) and cynodonts (*e. g., Thrinaxodon* [Fourie, 1974], *Procynosuchus* [Kemp, 1979], *Luangwa* [Kemp, 1980] and *Kayentatherium* [Sues, 1985]), and of the frontal sinuses in *Morganucodon* (see Kermack and others, 1981) and in *Lambdopsalis* indicates that this is a rather primitive feature.

However, both the functions and the homologies of the paranasal sinuses in the various groups just mentioned are far from certain. Kermack and others (1981) suggested that the frontal sinus in *Morganucodon* is analogous, not homologous, with the frontal sinus in eutherians "in that there is no connection between the sinus and the nasal chambers in *Morganucodon*" (p. 34-35). Even though there is a connection between the sinus and the nasal cavity as seen in *Lambdopsalis*, it is still difficult to judge if the frontal sinus of *Lambdopsalis* is an independent development.

Nasal Cavity

Description

Most of the matrix was removed from the nasal cavity in IVPP V7151.52. Much of the following description is based upon V7151.52, but some details are derived from the following specimens: IVPP V5429.3, V7151.5, V7151.84, and V7151.85.

The nasal cavity is surrounded by several bony elements. Its roof is formed by the internarial processes of the premaxillae, the nasals, and the anterior parts of the frontal plates of the frontals. Its lateral wall is formed anteriorly by the nasal process of the premaxilla and the ascending process of the maxilla, and posteriorly by the orbital process of the frontal. The nasal cavity is floored by the palatal processes of the premaxillae and maxillae, and by the horizontal processes of the palatines.

The nasal cavity is incompletely divided bilaterally by a bony vertical median septum. In its posterior part, the nasal cavity also is divided by a horizontal bony partition, here designated as "transverse lamina," into the olfactory chambers dorsally and the choanae ventrally. Hence, the nasal cavity consists of four compartments. Two dorsal compartments (*i. e.,* the olfactory chambers) are larger and separated partially from the olfactory fossa of the cranial cavity by medial extentions of orbital processes of the frontals. However, dorsal to this bony partition is a large, transversely elongated opening (the opening might have been subject to a little dorsoventral compression during fossilization), here dubbed as "olfactory opening." The olfactory opening connects the olfactory chambers of the nasal cavity with the olfactory fossa of the cranial cavity. In life, a cribriform plate occupying the olfactory opening would have been expected, but it is not preserved in specimens examined. Two ventral compartments, the choanae, are much smaller and end posteriorly at the internal nares. The internal nares are at the level of the posterior margin of the secondary palate and about at the posterior third of M2 in ventral view. The median septum of the choanae extends posteriorly beyond the level of the internal nares, though gradually decreasing its height during the course.

Even though various bony partitions exist within the nasal cavity, it communicates with the cranial cavity by the olfactory opening (or by foramina in the cribriform plate if such a plate existed). The nasal cavity communicates with the: oral cavity by the incisive foramina; nasopharynx by the choanae; and orbits by the nasolachrymal canals and the sphenopalatine foramina. The nasal cavity also has a connection with the frontal sinuses.

The specimen V5429.3 shows a transverse section of the nasal cavity at a level 3 mm anterior to the infraorbital foramen. Considerable bony remnants contained in the matrix can be observed from the section. These bony remnants are reminiscent of maxilloturbinals and are in the appropriate position for modern mammals.

The ridges and troughs on the inner surface of the nasal cavity have been described in association with the various individual bones surrounding it.

Discussion

The nasal cavity of *Lambdopsalis* as described above appears to be essentially typical of mammals, except for having an internarial bar. Until now, anatomical details of the nasal cavity in multituberculates have remained unknown, despite many publications on general cranial anatomy. Detailed knowledge of the nasal cavity of other groups of Mesozoic mammals also is virtually nonexistent. A possible exception to this generalization may be Kermack and others' (1981) study on the skull of *Morganucodon*. They devoted about a page and a half to the discussion of the nasal cavity, and their reconstruction of the nasal cavity in *Morganucodon* is essentially along typical mammalian lines. However, it should be pointed out that pertinent evidence to support their reconstruction is difficult to assemble from the available fragmentary materials. Therefore, as they admitted, the series of transverse sections through the nasal cavity of the reconstructed skull of *Morganucodon* (see Kermack and others, 1981, p. 123, Fig. 101) is largely conjectural.

In contrast, nasal cavities in some therapsids have been studied in detail (notably Kemp, 1969; see also Brink, 1960; Cox, 1959; Kemp, 1979, 1980; Tatarinov, 1963; and Watson, 1913). Because the secondary palate in therapsids has been considered to have allowed simultaneous mastication and respiration, along this line of reasoning, the structure of the nasal cavity in these mammal-like reptiles has been regarded as truly mammalian (Brink, 1957). These interpretations naturally led to speculation on the existence of endothermy in mammal-like reptiles and early mammals (Brink, 1957, 1980; and McNab, 1978). Interpretation of maintenance of endothermy in mammal-like reptiles implies high metabolic rates and, in turn, necessitates high nutrient demands. Thus, an endothermic physiology in mammal-like reptiles has been suggested as a major selective force in development of the secondary palate and the mammalian type of nasal structure (*e. g.*, Kemp, 1982; Moore, 1981). This seemingly circular approach has been questioned recently by a theoretical analysis of mechanical factors in the evolution of mammalian secondary palate (Thomason and Russell, 1986). They proposed that the secondary palate in mammal-like reptiles and mammals may perform an important mechanical function in resisting the forces of mastication and that mechanical factors may have played an even more crucial role in the development of the secondary palate. Nevertheless, it seems that the interpretations offered by the conventional wisdom and these two authors are not mutually exclusive. In fact, Thomason and Russell (1986, p. 212) pointed out that: perhaps the most plausible explanation ". . . is that the secondary palate and endothermy both arose in a lineage that was developing other compatible features, such as heterodont dentition and powerful, coordinated jaw muscles, resulting in a whole that was more than the sum of the parts."

Additionally, Bennett and Ruben (1986) agreed with and went beyond these authors by stating: "The primary function of the bony palate appears to be masticatory, as it serves as a platform for manipulation of the food by the tongue" (p. 211). Their arguments are that: (1) ". . . a bony secondary palate is present in the ectothermic crocodilians and teiid lizards, but is lacking in the endothermic birds" (*ibid.*, p. 211); and (2) in many living ectotherms, a fleshy partition, the soft palate, is equally effective in separating the ventilatory stream. However, Bennett and Ruben believed that the possession of complex turbinals provides strong evidence for interpretation of endothermy in therapsids.

Therefore, at least in some advanced cynodonts and the earliest mammals such as morganucodontids and multituberculates, it is reasonably safe to postulate that an endothermic physiology may have evolved. Detailed features of the nasal cavity of *Lambdopsalis* may be considered as added evidence to this assertion.

Median septum.—It is impossible to distinguish the specific contributions to the median septum of the nasal cavity from the various surrounding bony elements. However, the posterior extension of the median septum in the choanal region beyond the level of the internal nares suggests that the vomer may provide a substantial contribution. In the absence of evidence to the contrary, one may assume that the ventral part of the median septum receives contributions from the palatine processes of the premaxillae and maxillae, as well as the horizontal processes of the palatines. In mammals in general, the anterior expansion of the cranial cavity has brought the presphenoid in contact with the vomer, making the presphenoid contribute, at least to some extent, to the formation of the median septum. However, in many mammals, more important contribution to the median septum comes from the bone of the mesethemoid ossification (Moore, 1981). This may be also true of *Lambdopsalis*. Failure to distingush the vomer from the mesethemoid (if it existed at all), however, prevents further discussion on this matter for *Lambdopsalis*.

If the median septum of *Lambdopsalis* included a substantial contribution from the vomer, the vomer would have been a single bone occupying a similar position as in other mammals. Although homology of the mammalian vomer to vomers of mammal-like reptiles and modern reptiles has long been in debate (De Beer and Fell, 1936; Parrington and Westoll, 1940; Presley and Steel, 1978), it is probable that the single vomer in the early mammals is homologous to that of modern mammals.

The transverse lamina that separates the olfactory chambers from the choanae may be formed by ossifications from the vomer, maxillae, and palatines. However, the nature of this bony formation cannot be pursued further with available fossils.

Ethmoid.—In his excellent review of the mammalian skull, Moore (1981, p. 243) stated: "As compared with the typical reptilian condition, the new elements in the mammalian nasal region are the ethmoid and the repositioned vomer. The former element, in its fullest development, consists of perpendicular plate (with its superior projection, the crista galli), cribriform plate and right and left labyrinths." Stemming from Broom's (1926, 1927) studies of development of the sphenethmoid complex in various groups of modern mammals, this notion, as reemphasized above by Moore, has been popularized by Goodrich (1930) and De Beer (1937), and passed on for more than half a century. Broom's studies showed that the mesethmoid (*i. e.,* the perpendicular plate of the ethmoid) is present in some mammalian orders, but absent in others. He interpreted that the mesethmoid is a neomorph evolved within Mammalia, and thereby called those orders with a mesethmoid as Neotherida and those without it as Paleotherida.

In his study of serial sections of therapsid skulls, Olson (1944) described a mesethmoid in gorgonopsids, therocephalians and anomodonts. Therefore, Olson seriously questioned Broom's view, and considered the mesethmoid as a holdover from the therapsids to mammals; he deemed its absence in some mammalian orders as a secondary loss. However, Cox (1959) failed to detect a separate ossification as a mesethmoid in a dicynodontid anomodont (*Kingoria),* and implicitly doubted Olson's identification.

Nevertheless, Olson's interpretation of the presence of a mesethmoid in therapsids has been followed by many paleontologists (*e. g.,* Kemp, 1979, 1980; Kühne, 1956; and Tatarinov, 1963). These authors interpreted the presence of a mesethmoid in therocephalians and cynodonts, based upon either a ridge or a fossa on the internal surface of the frontal. In other words, no actual separate mesethmoid bone was found in specimens they described. The same is true for *Morganucodon* (see Kermack and others, 1981). In *Lambdopsalis*, even though there is a bony median septum, it is presently impossible to identify, or even recognize, individual bony elements. Therefore, I prefer to call the structure "median septum" in purely descriptive terms.

It should be noted that Olson's challenge to Broom's theme was unconvincing because it is difficult, if not impossible, to delimit the individual ossifications in the nasal region of therapsids. Roux (1947) pointed out the uncertainty of homology of individual ossifications of the nasal region even among modern mammalian groups in which prenatal stages of ossification have been carefully traced. Such difficulties seem to trivialize the importance that Broom attached to the presence versus absence of a mesethmoid in phylogenetic interpretation. In my opinion, it would not be surprising if the therapsids, the predecessors of mammals, indeed possessed the mesethmoid; but if they did, we cannot identify the bone from fossils.

Moreover, although Broom (1935) claimed there is no trace of a mesethmoid in monotremes (based upon a half-grown individual of *Ornithorhynchus* in the American Museum of Natural History), Jollie (1962, p. 46) pointed out that in *Ornithorhynchus* "There is a small perpendicular plate of the ethmoid fused to the anterodorsal end of the orbitosphenoid." Great caution, therefore, should be taken before putting much weight on phylogenetic implications of this poorly understood bone.

Interestingly, a similar interpretive dilemma has developed in dealing with presence versus absence of the cribriform plate in therapsids and early mammals. For example, Kemp (1969) noted that a small plate is present between the olfactory opening and the median septum in *Arctognathus* sp. (a Late Permian gorgonopsid) and thought that ". . . it is difficult to avoid the conclusion that this structure . . . appears to correspond to the mammalian cribriform plate" (p. 59). In the description of the nasal cavity of *Morganucodon*, Kermack and others (1981, p. 36) wrote: "The cribiform [sic] plate (cri. pl.) forms the posterior wall and separates the nasal chamber from the cranial cavity. The cribiform [sic] plate was attached to the transverse and oblique (obl. r. fr.) ridges of the frontal." However, careful reading of information provided elsewhere in the monograph indicates that no cribriform plate has been found in materials available to them.

Kielan-Jaworowska and others (1986) suggested a cribriform plate was present in *Nemegtbaatar*, based upon tiny bone remnants contained in matrix seen in appropriate sections of the skull.

In *Lambdopsalis*, though a cribriform plate might have been attached dorsally to the transverse ridge of the frontal, the plate itself has not been found. Admittedly, the chance for this delicate element to be preserved in a fossil (and, even if preserved, to be prepared out of the matrix) is slim.

Despite problems discussed above, the cribriform plate is a constant structure among modern terrestrial mammals. Even in Broom's Paleotherida, which he presumed to have lacked mesethmoids, there consistently exists a cribriform plate, except in some aquatic forms such as toothed whales and platypus. Therefore, it is indeed tempting to postulate that the cribriform plate evolved in the earliest mammals which, supposedly as noctural animals, might owe their early success to olfactory acuity. However, my point is that although we can speculate about its possible (or even probable) existence, we should clearly separate speculation from repeatable observation.

Turbinals.—In modern reptiles such as lizards and snakes, "A large fold or turbinal, slightly rolled on itself, arises from the outer wall of the inner nasal chamber, and extends far into its lumen" (W. N. Parker, 1907, p. 265). The turbinal in reptiles is often referred as the concha, to distinguish it from the more complex turbinals in mammals. In crocodiles, the concha is subdivided further into preconcha, middle concha, and postconcha (Parsons, 1959). Reptilian conchae are cartilaginous, and mammalian turbinals are mostly ossified (though seldom fossilized). In life they are attached to the bony ridge(s) on the given bone(s). For example, Oelrich (1956, Fig. 26) described and illustrated the conchal ridge on the ventral surface of the nasal in *Ctenosaura pectinata.* Similar but more complex ridge systems are observable on the inner sides of bones surrounding the nasal cavity in any dissected modern mammalian skull. These led Watson (1913, 1951) to believe that longitudinal ridges on the ventral surface of the nasal bones in cynodonts (*e. g., Diademodon* and *Nythosaurus*) indicated presence of turbinals, probably similar to those of modern mammals. This view has been followed by virtually all later workers in the field (*e. g.*, Brink, 1957, 1980; Kemp, 1969, 1979, 1982; Kermack and others, 1981; Kühne, 1956; and Van Valen, 1959).

The complex ridge system present on the ventral surfaces of the nasal, frontal, and maxilla in *Lambdopsalis* resembles arrangements of ridges seen in *Morganucodon* and in modern mammals. This seems to further strengthen Watson's postulation, though the cautions I made earlier in discussions of the mesethmoid and cribriform plate also apply here.

Although much has been written about implications of the presumed presence of a mammalian pattern of turbinals in therapsids and early mammals, little attention has been paid to conflicting evidence between paleontology and embryology concerning the phylogenetic polarity of ethmoturbinals and maxilloturbinals. Moreover, the contradictory views within each of the two subdisciplines also have been overlooked.

A predominant view among neontologists is that reptilian conchae correspond to the maxilloturbinals of mammals, based upon the appearance of the maxilloturbinal earlier than the ethmoturbinal in mammalian ontogeny (Allison, 1953; De Beer, 1937; Kingsley, 1926; Moore, 1981; and Parsons, 1959). De Beer (1937) considered the preconcha of crocodiles to be homologous with mammalian maxilloturbinals, and suggested that the postconcha of crocodiles "is probably an independent development, not homologous with the mammalian ethmoturbinals" (p. 397). This implies that the ethmoturbinals are peculiar to mammals. Although De Beer (1937, p. 397) admitted that recognition of homology of turbinals in different vertebrate classes is "a matter of great difficulty," his opinion has been taken more conclusively than he originally intended (Allison, 1953; Moore, 1981).

Quite in contrast, a prevalent opinion among paleontologists was aptly expressed by Kermack and others (1981, p. 37) as follows: ". . . that the ethmoturbinals of mammals are conservative and go back with little change to the cynodonts of the Lower Trias: the maxilloturbinals may be a later, purely mammalian development." This belief is derived directly from Watson's work mentioned above. Watson (1913) described the endonasal casts (of *Diademodon* and *Nythosaurus*) in ironstone, and postulated presence of nasoturbinals and ethmoturbinals but absence of maxilloturbinals in these taxa. As Kermack and others (1981) also mentioned, Kühne (1956, p. 26) stated that *Oligokyphus* ". . . shows conclusively that there are no ossified maxilloturbinals." This paleontological inference was made more credible by Watson's functional consideration of maxilloturbinals. Watson (1953) pointed out that the maxilloturbinals in mammals provide an air-conditioning plant, which warms and humidifies the inspired air. It follows, therefore, that maxilloturbinals are neomorphic, clearly associated with the endothermic physiology of mammals.

Interestingly, Watson (1953, p. 125) also stated: ". . . the development of the naso- and maxilloturbinals in cynodonts suggests that, like mammals, they were warm blooded." However, he did not specify in which taxon (or taxa) in cynodonts the maxilloturbinals were believed to have existed. Nevertheless, the implication still stands that the maxilloturbinals may be a later (even though, in this case, not necessarily purely mammalian) development.

This otherwise straightforward discrepancy between paleontological and neontological evidence is further complicated by different opinions within each profession. While admitting that the reptilian concha is represented by the maxilloturbinals in mammals and that the ethmoturbinals are peculiar to mammals, W. N. Parker (1897, 1907) pointed out that ontogenetically the maxilloturbinals develop later than the ethmoturbinals. Jollie (1968, p. 274) also recorded that in the new-born specimen of *Manis javanica*, ". . . ossification of the maxillary turbinal has just begun," when the ethmoturbinals have already ossified. Moreover, in his study of the cranial development of Ethiopian "insectivores," Roux (1947) found: (1) that the first primary ethmoturbinals occupy the main ossification center in the ethmoid region "which appears relatively early and which is relatively constant in position" (p. 377); and (2) that in embryonic materials of *Suncus orangiae*, the first distinguishable ossification appears in the 18.3 mm embryo for ethmoturbinals, in the 22 mm embryo for maxilloturbinals, and in the 25 mm embryo for nasoturbinals (see Roux, 1947, p. 230-231). These observations imply that maxilloturbinals are phylogenetically older but ontogenetically later in development than ethmoturbinals. Considered in terms of Von Baer's Law, this implication is unusual. If true, it would throw the cladistic view of ontogeny as an important criterion in determining phylogenetic polarity into shadow.

Among paleontologists, on the other hand, Kemp (1969) interpreted the pair of turbinals in the maxillary sinus in *Arctognathus* sp. as the sinus turbinals, homologous with maxilloturbinals in mammals. This would imply that the maxilloturbinals could have arisen early among therapsids, when an endothermic physiology may not yet have evolved. However, Kemp (1969, p. 63) did point out that ". . . it would equally be true that all these merely represent suitable sites for turbinal development, independently in both groups."

Confronted with these conflicting lines of evidence and points of view, firm phylogenetic conclusions cannot be justified at present. For instance, although the turbinals in adult mammals are named topographically after the particular bone to which they are attached, their ontogenesis and homologies across various groups are far from clear. The nasoturbinals are considered by some authors (*e. g.*, Allison, 1953; Moore, 1981; and Paulli, 1900a) as the first turbinal elements of the ethmoturbinals because they usually ossify in part from the ethmoid. Others (*e. g.*, W. N. Parker, 1897) considered that the nasoturbinals should be spoken of as *sui generis* because they no longer possess olfactory epithelium, as the ethmoturbinals usually do. Still others (*e. g.*, Kohncke, 1985) regarded the nasoturbinals more closely associated with the maxilloturbinals than with other turbinals because they both are carried by the elongated pars anterior in the chondrocranium.

Therefore, further information is needed both from developmental and paleontological studies before the phylogenetic polarity of various turbinals can be determined. Parsons (1959, p. 182) once said about the turbinals in advanced therapsids: ". . . the presence of an essentially mammalian condition in forms definitely ancestral to mammals and very distant from any other recent group is hardly surprising, and tells nothing about the origin of that condition." It seems premature, however, to deny paleontological contributions to phylogenetic inferences. For example, the developmental "evidence" on which the reptilian concha-mammalian maxilloturbinals homology was based can be doubted. A similar situation occurred in determination of homology of the mammalian vomer with the reptilian parasphenoid as based upon developmental studies. It turns out, in this latter case, that most embryologists now have agreed that greater weight should be attached to the paleontological evidence; the mammalian vomer should be considered homologous to the fused vomers of cynodonts (Moore, 1981; Presley and Steel, 1978).

Hence, paleontologists cannot yet be left out of the high table (a metaphor used by Maynard Smith, 1984), contrary to some cladists' claim that fossils cannot provide a way of determining phylogenetic polarity. Also contrary to the belief of some paleontologists nowadays, evidence drawn from developmental studies for assessing the phylogenetic polarity of a character is not necessarily superior to that drawn from paleontology. The contradictory evidence and views derived from the various original researches in developmental biology just discussed illustrate the frustrations and uncertainties that embryologists have always had, just as their paleontologist counterparts have suffered from the ambiguities of their evidence. As to these frustrations and uncertainties, Jollie (1968, p. 285-286) said: "The cross the comparative anatomist must bear is an extensive and conflicting literature — now something more than 100 years old . . . . Judicious, and perhaps lucky, selection is the key to use."

Parietal (Figs. 1, 4, 7, 9-12, 17)

Description

The parietal is shown well in the following specimens: IVPP V5429.4, V5429.14-16, V7151.50, V7151.53, V7151.55, V7151.77, V7151.79, V7151.87, and V7151.90.

The parietal is an extensive bone, forming the larger, posterior part of the roof of the cranial cavity. It has a long anterior process, dorsally overlapping the frontal plate and extending forward to contact the tip of the posterior process of the nasal. The dorsal surface of the parietal is smooth and convex (more strongly convex transversely than anteroposteriorly). Both sagittal crest and lambdoidal crest are well developed, and they are strongest at their junction. In dorsal view, the suture between the left and right parietals is clearly visible. The same is true for the suture between the parietals and occipital. The parietal overlaps the frontal anteriorly, and is overlapped by the occipital posteriorly and by the alisphenoid and the anterior lamina of the petrosal as well as the petrosal laterally. The ventrally directed descending plate of the parietal is not well developed, and thus does not contribute much to the formation of the lateral wall of the braincase.

There is no indication of an interparietal bone.

At least two striking features can be observed on the ventral surface of the parietal: (1) a deep median groove is continuous with that on the ventral surface of the frontal, but terminates at the apex of an anteriorly convex transverse ridge; and (2) posterior to the transverse ridge, the parietal is greatly thickened to form a swelling. The thickest part of the swelling is in the sutural zone between the parietal and occipital. The broken part of the parietal seen from dorsal view in IVPP V7151.90 shows a system of pneumatic chambers within the swelling, here designated as the parietal sinus.

Anterior to the transverse ridge, there appear to be some fine grooves on the ventral surface of the parietal; but posterior to the ridge, the surface of the swelling is quite smooth. The suture between the left and right parietals cannot be seen on the ventral surface.

There are two foramina seen externally along the lateral edge of the parietal. The anterior one, here designated as postorbital foramen, is along the lateral edge of the long anterior process of the parietal, at the level of the anterior end of the sagittal crest and posterior to the postorbital process. The posterior one, here termed as the foramen of the ascending canal, lies within the suture at the triple junction of the parietal, anterior lamina of petrosal, and squamosal, just above the subarcuate fossa of the petrosal. On the ventral surface of the parietal, a longitudinal groove connects these two foramina, here designated as the internal parietal groove.

Discussion

The parietal is a dermal bone covering the cranial roof in the temporal region. The overall shape of the parietal of *Lambdopsalis* is more similar to that of the North American Paleocene multituberculates (especially *Taeniolabis*) than to that of the Late Cretaceous Mongolian taxa as reconstructed by Kielan-Jaworowska (1971). In *Kamptobaatar*, she originally reconstructed an extensive contribution from the descending plate of the parietal to the lateral wall of the braincase (*ibid.*). Clemens and Kielan-Jaworowska (1979) emended her original reconstruction, and showed that the parietal barely extends onto the lateral wall of the braincase either in *Kamptobaatar* or *Nemegtbaatar*. The clear suture between the parietal and the temporal bony elements seen in *Lambdopsalis* demonstrates that, indeed, the parietal contributes little to the formation of the side wall of the braincase. The same is true of *Taeniolabis* (see Simpson, 1937). Therefore, one may suspect that the conjectural line of the suture between the parietal and squamosal in *Ptilodus* as drawn by Simpson (1937) should be more dorsally situated.

Like *Ptilodus* and *Taeniolabis*, *Lambdopsalis* also has a relatively long anterior process of the parietal, laterally overlapping the frontal. However, the anterior

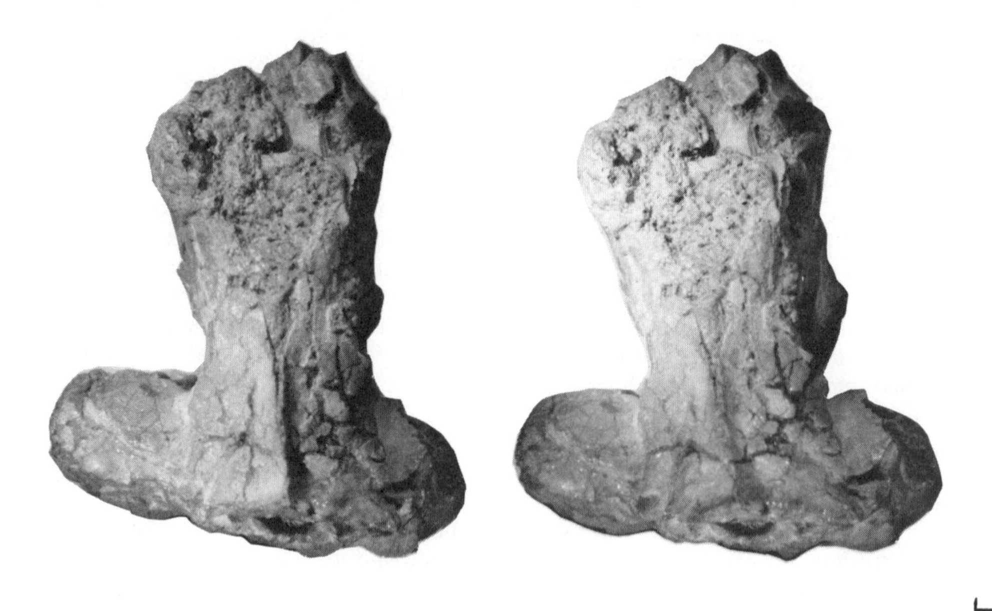

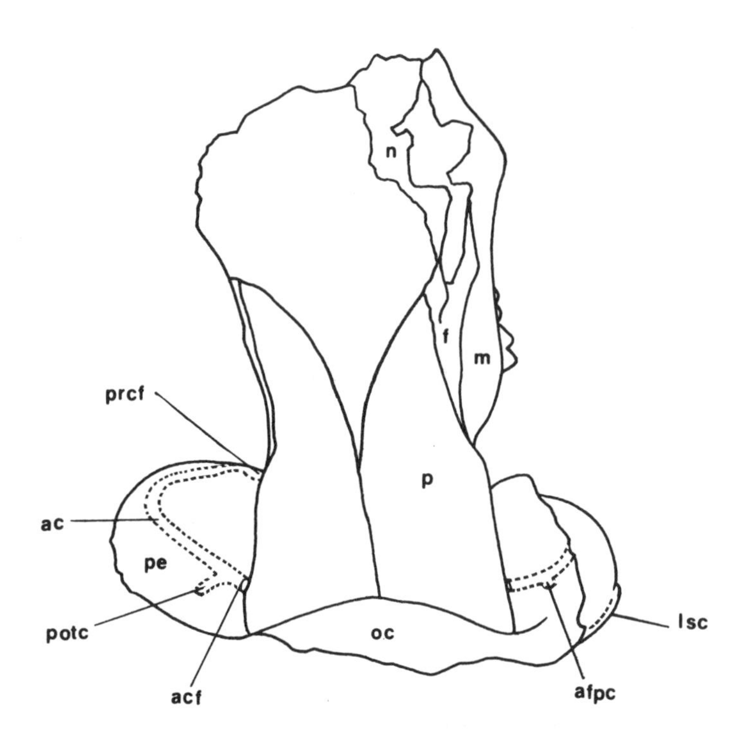

Figure 10. ***Lambdopsalis bulla*****, V7151.77, dorsal stereoscopic view of skull. Anterior to top.**

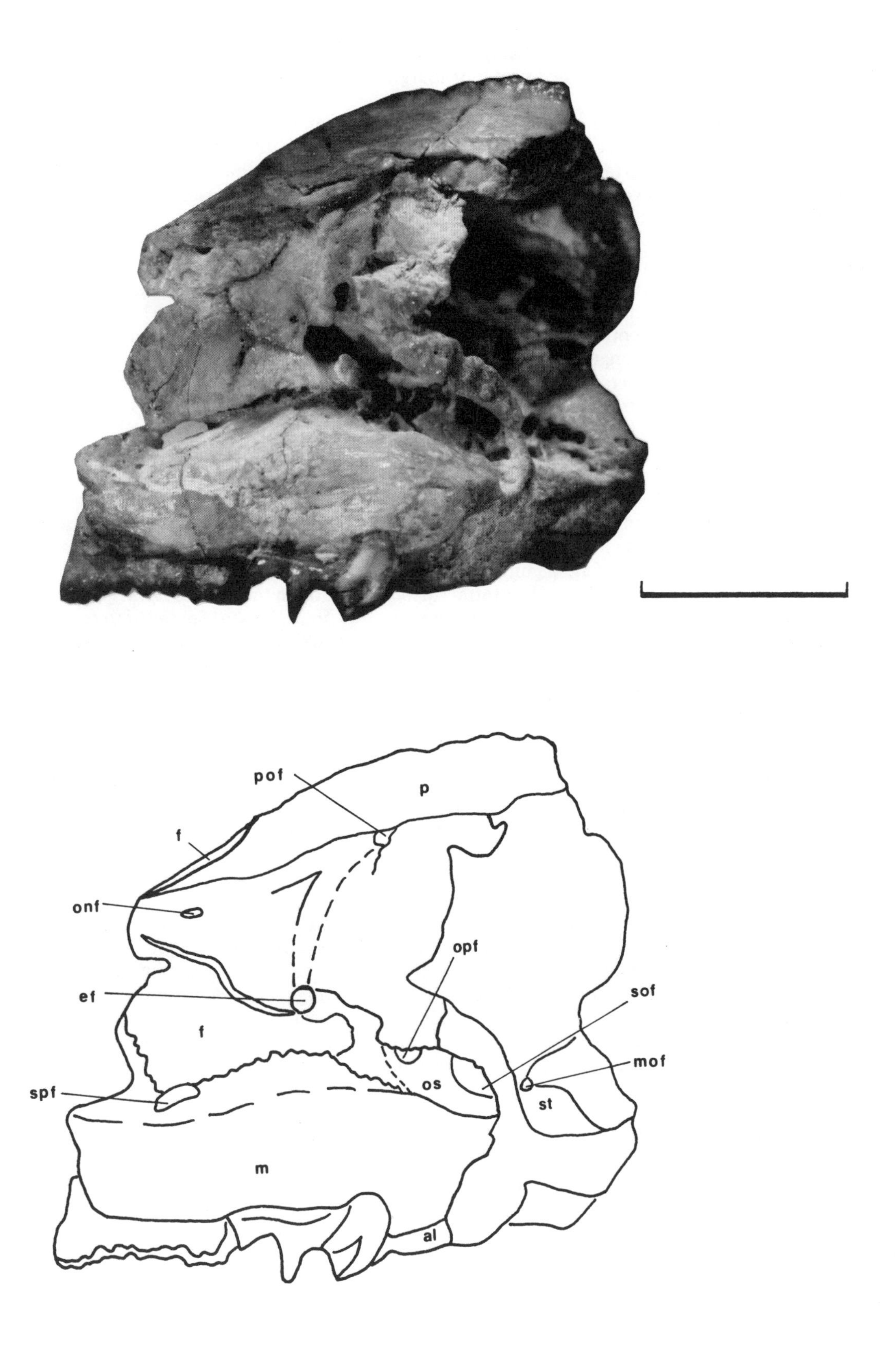

Figure 11. ***Lambdopsalis bulla*, V7151.52, lateral view of fragment of skull. Anterior to left.**

process of the parietal in the Late Cretaceous Mongolian multituberculates is much shorter, probably due to the extensiveness and transversely-aligned posterior margin of the frontal.

Median groove, transverse ridge, and swelling.—Kermack and others (1981) described a median, longitudinal ridge on the ventral surface of the parietal in *Morganucodon*, and they believed it served as a dorsal bony separation between the cerebral hemispheres. In contrast, a median, longitudinal groove occupies the similar position in *Lambdopsalis*. This distinct groove is present in an expected position along which the superior sagittal sinus runs in many living mammals. Even though there is no such groove on the ventral surface of the parietal in the Late Cretaceous Mongolian multituberculates, Kielan-Jaworowska and others (1986) still included this venous sinus in their reconstruction, based upon its consistent occurrence in recent mammals. It is reasonable to assume that the median groove in *Lambdopsalis* may have contained the superior sagittal sinus. Nevertheless, the significance of this difference in topographic relief beneath the sagittal suture between *Lambdopsalis* and *Morganucodon* is not clear. However, it should be stressed that in a majority of living mammals there is no corresponding osteological feature on the ventral surface of the parietal to match with the median fissure between the cerebral hemispheres.

Patterson and Olson (1961) described a triangular swelling or projection in the posterior part of the ventral surface of the parietal in *Sinoconodon rigneyi*, which is reminiscent of the swelling here described in *Lambdopsalis*. They suggested that the swelling in *Sinoconodon* formed a major separation between cerebrum and cerebellum, and was analogous to an ossified tentorium in some therians. Kermack and others (1981) described a transverse ridge in *Morganucodon*, which is reminiscent of the transverse ridge here described in *Lambdopsalis*. However, Kermack and others (1981) equated the transverse ridge of *Morganucodon* with the triangular swelling of *Sinoconodon*. The fact that both the transverse ridge and swelling exist in *Lambdopsalis* in comparable positions may cast doubts both on Kermack and others' interpretation of homology of the transverse ridge of *Morganucodon* with the swelling of *Sinoconodon*, and on Patterson and Olson's interpretation of the swelling in *Sinoconodon* being an ossified tentorium. My arguments are as follows:

1. Morphologically, the transverse ridge of *Morganucodon* is dissimilar to the swelling of *Sinoconodon* (the latter is so impressive that Patterson and Olson [1961, p. 152] called it "a truly remarkable structure"). If I were forced to speculate about the homologies of these structures in the taxa under discussion, I would consider the transverse ridge of *Morganucodon* equivalent to the transverse ridge of *Lambdopsalis* and the swelling of *Sinoconodon* equivalent to the swelling of *Lambdopsalis*. The transverse ridge of *Morganucodon* cannot be equated with the swelling of *Sinoconodon*.

2. The swelling in *Lambdopsalis* contains a system of pneumatic chambers that are sinus-like. This well-developed parietal sinus system might be related to the bone conduction hearing mechanism of *Lambdopsalis*, but this problem will be considered later. However, it is not known whether the swelling in *Sinoconodon* also contains a sinus-like structure. Patterson and Olson (1961) did note that the swelling of *Sinoconodon* is far more massive than a typical ossified tentorium.

3. Posterior to the swelling, there is a very narrow trough in *Sinoconodon*, which was designated as "cerebellar cavity" by Patterson and Olson (1961). Similarly, a narrow trough is also present in *Morganucodon* (posterior to the transverse ridge); it was designated as "cerebellum trough?" by Kermack and others (1981). However, Kermack and others noted the narrowness of both "cerebellar cavity" of *Sinoconodon* and "cerebellum trough?" of *Morganucodon*, and preferred to consider them as being without known function. It also should be noted that the width of the so-called cerebellar cavity in *Sinoconodon* is about 1 mm, and that of cerebellum trough? in *Morganucodon* about 1.5 mm. The cavity or trough would be too narrow, as Kermack and others realized, to accommodate the dorsal exposure of the cerebellum (see also Kielan-Jaworowska, 1986).

4. The serial sections through the appropriate position in the skull of *Nemegtbaatar* (see Kielan-Jaworowska and others, 1986) failed to reveal any transverse ridge and swelling. In addition, as I argued earlier, the ventral surface of the parietal in many recent mammals is nearly featureless. Therefore, although these structures seem to occur in a few taxa concerned, they are by no means constant either in early mammals or in extant mammals.

Admittedly, we understand neither the functions nor the taxonomic significances of those structures at this point, and further speculation seems not to be worthwhile.

Postorbital foramen, foramen of ascending canal, and internal parietal groove.—Kielan-Jaworowska and others (1986) described the postorbital foramen in *Nemegtbaatar* as a foramen in the parietal at the posterior end of the orbital groove, just below or within the postorbital process, and interpreted the foramen as transmitting blood vessels from the internal parietal groove to the orbit. The postorbital foramen in *Lambdopsalis* is similar to this, both topographically and, presumably, functionally. The only differences between the two genera are that: (1) the postorbital foramen in *Lambdopsalis* is at the triple-junction of the parietal, frontal, and alisphenoid, rather than within the parietal as in *Nemegtbaatar*; and (2) the postorbital foramen in *Lambdopsalis* is posterior to, rather than below or within, the postorbital process. This last difference, in turn, is related to the possibility that the postorbital process may not be homologous. As I suggested earlier, the postorbital process in the Late Cretaceous Mongolian taxa as reconstructed by Kielan-Jaworowska and others (1986) may not be equivalent to the postorbital process of other mammals. In addition, the postorbital foramen in *Lambdopsalis* is more anteriorly situated than that in *Nemegtbaatar*. Consequently, the orbital groove (*i. e.*, the groove extending anteroventrally from the postorbital foramen to the ethmoid foramen) in *Lambdopsalis* is proportionally shorter than that seen in *Nemegtbaatar*.

The foramen of the ascending canal in *Lambdopsalis* is similar to that in *Nemegtbaatar* as described by Kielan-Jaworowska and others (1986). The only difference is that the foramen in *Lambdopsalis* is within the suture at the triple junction of the parietal, the anterior lamina of petrosal, and the squamosal; the foramen in *Nemegtbaatar* is within the suture between the anterior lamina and the squamosal. In both *Lambdopsalis* and *Nemegtbaatar*, the foramen opens the intramural ascending canal into the cranial cavity (the ascending canal will be described and discussed in the section entitled "Petrosal and Auditory Region").

Kielan-Jaworowska and others (1986) introduced the term "internal parietal groove" in describing the segment of the groove that runs more or less anteriorly from the foramen of the ascending canal to the postorbital foramen in the Late Cretaceous Mongolian multituberculates. This term is applicable to the similar structure here described in *Lambdopsalis*. They also discussed the nature of the vascular vessels carried by the internal parietal groove, and equated the groove with the sinus canal of cynodonts and also with the groove for the "vena temporo-orbitalis" of *Morganucodon*. While I save speculation about the vascular vessels for later discussion, I now consider the problem of recognition of homologies of the internal parietal groove in multituberculates with comparable structures in cynodonts and *Morganucodon*.

W. K. Parker (1885) described an intracranial "sinus canal" in embryonic insectivores, which is topographically nearly identical to the internal parietal groove in multituberculates as described here and by Kielan-Jaworowska and others (1986). Watson (1911) suggested homology of Parker's "sinus canal" of insectivores with his "sinus canal" of *Diademodon*. This usage has been followed in essentials by all later workers on therapsids. The blood vessels carried by the "sinus canal" of therapsids have been discussed by Cox (1959) for *Kingoria,* Crompton (1964) and Kühne (1956) for *Oligokyphus*, Fourie (1974) for *Thrinaxodon*, Hopson (1964) for *Bienotherium*, and Kemp (1979) for *Procynosuchus*. Despite some differences on certain details of the interpretations among these authors, almost all of them agree: (1) that the groove occupies similar anatomical position, placed extracranially along the suture zone between the parietal and epipterygoid plus prootic. [Kemp (1979, p. 104), however, pointed out that the groove in *Procynosuchus* ". . . lies well above the line of the suture of the parietal with the epipterygoid and the prootic." Nevertheless, it was implicitly suggested that the sinus canal is a homologous structure among those taxa]; and (2) that the sinus canal primarily contained a head vein rather than an artery.

Although homology of the sinus canal among various therapsid taxa has not yet been questioned, both its homology with the sinus canal in insectivores and its primary venous affinity were challenged recently (see respectively Kermack and others, 1981; and Kielan-Jaworowska and others, 1984, 1986). Kermack and others (1981, p. 90) pointed out: "Parker describes a sinus canal in a number of insectivores but this can have nothing whatsoever to do with the 'sinus canal' of Watson. Parker's canal is a groove on the inside of the cranial wall, while Watson's 'sinus canal' is a groove on the external surface of the bone." They also suggested a presumed difference in the sinus' vascular connection with different venous vessels between insectivores and cynodonts, and therefore concluded (*ibid.*, p. 90): "Watson (1911) was here incorrect in his homology." However, it is interesting to note that, in the same monograph, Kermack and others described and illustrated the "sinus canal" (their groove for the vena temporo-orbitalis) as being intracranial in *Morganucodon*, and yet equated it with the extracranial sinus canal of cynodonts.

Kielan-Jaworowska and others (1986) pointed out a similar flaw in Kermack and others' argument on this issue. Kielan-Jaworowska and others further demonstrated that blood vessels in the so-called "sinus canal" may have become either intracranial or extracranial without migrating through any tissue during the developmental remodeling of neighboring bones. Furthermore, in his study of the development of mammalian dural venous sinuses, Butler (1967, p. 33) stated: "Morphologically, as shown by Sutton (1888), both the dural venous sinuses and the external jugular venous system are extracranial in so far as they are situated outside the dura mater. The chondrocranium and the dermocranium, however, develop in between the dural venous sinuses and the external jugular vein system (Butler, 1957). Thus, in the adult mammal, the bony skull wall separates the two venous systems, and therefore the dural venous sinuses are topographically intracranial whereas the external jugular venous system is extracranial."

This shows clearly how the originally extracranial sinuses (remember the vessels contained in the "sinus canal" are conventionally considered as part of the dural venous system) become topographically intracranial during morphogenesis. Apparently, these authors' statements trivialized this topographic difference of the sinus canal between cynodonts and mammals, and favored their probable homologous nature.

Nevertheless, even though the argument of Butler (1967) and Kielan-Jaworowska and others (1986) seems convincing, a strong case still can be made from the fact that the so-called sinus canal in cynodonts lies extracranially, strikingly similar to that seen in modern reptiles (O'Donoghue, 1920), whereas the internal parietal groove (homologous to "sinus canal") in *Morganucodon, Lambdopsalis,* and a variety of the Late Cretaceous Mongolian multituberculates lies intracranially as in modern mammals. It is tempting to suggest that this topographic change in vascular route through the developmental remodeling of the neighboring bony elements may have occurred during the therapsid/mammal transition. This would be an example parallel to the topographic change of the cavum epiptericum during this transition (also from originally extracranial position changing into intracranial position). Similarly, this should be considered valuable at least for taxonomic purposes.

Furthermore, although Kielan-Jaworowska and others (1984; 1986) questioned the primary venous affinity of the blood vessels contained in the sinus canal, their challenge does not damage the above argument. Even if the sinus canal in cynodonts and the internal parietal groove in early mammals contained primarily arteries, the arteries, with companion veins, also may have undergone the topographic change without shifting through any tissue during the same process of the morphogenesis. However, the vascular nature of the internal parietal groove (and the "sinus canal" of cynodonts) will be further explored in connection with the later discussion of the cranial vasculature in *Lambdopsalis*.

"Jugal"

Description

No complete zygomatic arch has been found preserved in the available collection of *Lambdopsalis*. The zygomatic processes of both the maxilla and squamosal are partly preserved; however, they are too incomplete to show any suture or facet that might suggest either presence or absence of a jugal bone.

Discussion

It is of interest to note that considerably more attention has been paid to the jugal in multituberculates than to that in any other mammalian group. Gidley (1909) originally described the jugal of *Ptilodus* (USNM 6076) as such: "The malar extends backward to the glenoid surface, and apparently joins the lachrymal bone anteriorly as in the living marsupials. The anterior extension of the malar, however, can not be made out with certainty, owing to the almost complete obliteration of suture lines in this region" (p. 619). It should be pointed out that Gidley reconstructed the jugal of *Ptilodus* as in the living marsupials because he (*ibid.*, p. 625-626) believed that ". . . the Allotheria represent an extinct group of multituberculate eutherian mammals closely related with but not ancestral to the Diprotodont division of the Marsupialia."

Broom (1914) reexamined Gidley's *Ptilodus* skull, and admitted that the exact limits of the anterior process of the squamosal, jugal, and posterior process of maxilla in the zygomatic arch ". . . cannot be seen with certainty as both arches are crushed and imperfect" (p. 123). Yet he expressed that the maxillary and squamosal contributions to the arch are both large, with possibly a small jugal in between. At any rate, he did not agree with Gidley on the posterior extension of the jugal into the glenoid facet.

In the same paper, Broom also reconstructed a jugal in *Taeniolabis taoensis* (then known as *Polymastodon taoensis*). The jugal was reconstructed as a small one resting on the anterodorsal border of the zygomatic arch, which is formed mainly by the processes of the maxilla and squamosal. This reconstruction of the zygomatic arch of *Taeniolabis* was largely based upon *Ornithorhynchus*. It should be noted that the zygomatic arch in the *Taeniolabis* specimen studied is also badly crushed, and that the jugal cannot be discerned. About this, however, Broom (*ibid.*, p. 128) stated: "Though imperfect, it must be practically as I have restored it." He even went further to suggest that the jugal of *Ptilodus* is possibly only on the dorsal side of the zygomatic arch as in *Taeniolabis*, unlike what he previously represented as separating the maxilla from squamosal. Subsequently, Broom (1914) used this character, along with a few other equally doubtful ones, to argue for a close relationship between multituberculates and monotremes. Circular reasoning along this line in Broom's paper is obvious.

Simpson (1937) was aware of this problem. With addition of a better specimen of *Ptilodus* (AMNH 35490), he reinvestigated all the materials that Gidley (1909) and Broom (1914) previously studied, plus a Late Cretaceous Mongolian multituberculate, *Djadochtatherium*. Even with that better specimen, Simpson (1937) was unable to solve the problem. He described the whole anterior root of the zygomatic arch in AMNH 35490 as being formed by the maxilla, and the zygomatic process of the maxilla extending posteroventrally almost to the glenoid surface. Simpson (*ibid.*, p. 743-744) also stated: "Gidley and Broom both considered this posterior extension as belonging to the jugal . . . but in my better specimens it seems almost certain that there is no suture between this and the maxilla." However, as I noted above, Broom did not consider that posterior extension as part of the jugal, contrary to Simpson's statement. Nevertheless, Simpson did not exclude the possibility of a small jugal above the zygomatic arch in multituberculates. Moreover, Simpson (1937, p. 744) said: ". . . we do not know what really happened to the jugal in the multituberculates." Simpson further stressed that even if the zygomatic arch and jugal of multituberculates are monotreme-like, they are rather superficial in resemblence, and probably are not suggestive of phylogenetic affinity.

Kermack and Kielan-Jaworowska (1971) and Kielan-Jaworowska (1971) revived Broom's idea of multituberculates being closely related to monotremes. Again the nature of the jugal was taken as substantial evidence. However, unlike Broom, these authors claimed both multituberculates and monotremes lack a jugal bone. Whereas the absence of a jugal is only true of tachyglossid monotremes (Griffiths, 1968, 1978), lack of a jugal in multituberculates was neither confirmed nor disproved by fossils at that time. When Kielan-Jaworowska (1971) described a skull of *Sloanbaatar* with a complete zygomatic arch, she clearly pointed out: "The suture between the maxilla and squamosal on the zygomatic arch is not discernible" (p. 18). However, she believed that the jugal is absent, and that the zygomatic arch is formed only by the maxilla and squamosal. This clearly demonstrates how one's preconceived idea of phylogeny of a certain group can influence one's assessment of certain characters in that group (rather than vice versa).

Almost a century ago, Slade (1895, p. 50) complained: "It is difficult to explain why that portion of the mammalian cranium which presents so prominent and striking a feature, even to the most careless observer, as does the jugal or zygomatic arch, should not have been considered worthy of more extended scientific notice than it has received." Today one might as well complain why so much attention has been paid to the zygomatic arch, especially the "jugal", in multituberculates whereas definite existence of the jugal in them was not convincingly documented until very recently.

No evidence from study of *Lambdopsalis* can solve this problem. However, Kielan-Jaworowska and others (1986) reported (in their "Notes added in proof") Hopson's discovery of a slender jugal in *Ptilodus* (the same specimen originally studied by Simpson, 1937), and the subsequent discovery of a slender jugal in several specimens of the Late Cretaceous Mongolian taeniolabidoid genera by Kielan-Jaworowska. In all cases, the jugal is internal to the zygomatic arch, and not visible in lateral view.

With these findings, it may be instructive to ask what kind of phylogenetic indication the characteristics of a jugal might bear on mammalian phylogeny. According to Hogben (1919), the primitive condition for the jugal is probably seen in most living marsupials and less specialized eutherians, in which the jugal arises from the anteroventral corner of the orbit and extends posteriorly to the glenoid fossa. Frequently, however, the zygomatic processes of the maxilla and squamosal encroach upon the jugal, with the reduction of the latter to varying degrees in many mammalian groups. By this interpretation, the diminution and loss of the jugal in monotremes are derived conditions. The same is probably true of edentates, lagomorphs, chiropterans, erinaceomorphs, and fossorial talpid insectivores. The reduction or absence of a jugal in these taxa also may be secondary, and related to reduction of its functional complex (the zygomatic arch) as a whole. Therefore, even assuming existence of some multituberculates with a reduced jugal or without a jugal at all, it does not necessarily denote their phylogenetic affinities with either monotremes or talpid insectivores. They certainly have acquired these characters quite independently. Additionally, multituberculates probably were nearly as diverse as rodents, and thus the jugal in multituberculates might be as variable as that in Recent rodents. It is advisable not to put too much phylogenetic weight on characteristics of the jugal alone. The fact that platypus has but echidnas do not have a jugal suggests that variation of the character does exist even within one of the least diverse of mammalian groups.

Squamosal (Figs. 12, 17, 18)

Description

No complete squamosal is known in the collection. However, most parts are well preserved in the following specimens: IVPP V7151.64, V7151.74, V7151.75, and V7151.88-90. The contact surface for the squamosal can be observed in many isolated petrosal bones in the collection. Thus the confines of the squamosal are clearly defined, except for its anterior limit in the zygomatic process.

The squamosal is in sutural contact with the parietal, the anterior lamina of the petrosal, and the main body of the petrosal. The zygomatic process of the squamosal ought to be in sutural contact either with the "jugal" (if there was one) or with the zygomatic process of the maxilla. This is impossible to determine due to incompleteness of the zygomatic arch.

Much of the squamosal laterally overlaps the anterolateral part of the petrosal. The external surface of the squamosal is convex and smooth. In contrast, the internal surface is concave and rough. The squamosal can be divided arbitrarily into a body of squamosal, a dorsal flange, and a zygomatic process.

The dorsal flange of the squamosal extends dorsomedially. Its anterior tip articulates with (and overlaps) the parietal dorsally and the anterior lamina of the petrosal ventrally. The dorsal flange is thin and tapers dorsomedially. The triple junction of the dorsal flange, parietal, and the anterior lamina is situated at about the convergent point of the lateral and posterior margins of the braincase; therefore, the squamosal makes almost no contribution to either the cranial roof or the side wall of the braincase. The dorsal flange terminates posteriorly at a ridge that is known as the lambdoidal crest of other mammals. Anteriorly, the dorsal flange ends at the anterior rim of the ascending canal, which is a deep groove on the external surface of the petrosal covered laterally by the squamosal. At about the middle point of the dorsolateral edge of the sutural zone between the dorsal flange and the petrosal, the ascending canal gives off a posteroventrally directed canal, here designated as the post-temporal canal. In other words, the anterior opening of the post-temporal canal in *Lambdopsalis* is covered by the dorsal flange of the squamosal, and the post-temporal canal is mainly intramural (*i. e.,* within the petrosal).

The body of the squamosal is fishscale-like, and laterally covers much of the anterolateral surface of the petrosal. The zygomatic process of the squamosal arises from the body of squamosal, and extends anteriorly. The best-preserved zygomatic process of the squamosal is broken at a position just beyond the anterolateral rim of the glenoid fossa, and thus its anterior limit cannot be determined. The glenoid fossa is oval, with its anterior part narrower than its posterior part. The long axis of the glenoid fossa is oriented anteroposteriorly. The glenoid fossa is concave, with slight lateral and medial rims. The medial rim forms a ridge which slopes down ventromedially to terminate at the suture between the ventral margin of the squamosal and the petrosal. The posteroventral part of the squamosal overlies much of the mastoid part of the petrosal.

Discussion

The squamosal is a dermal bone, which mainly covers the quadrate in the lower vertebrates, and becomes a dermal shield of the auditory capsule in the higher forms, particularly in mammals (Thyng, 1906). De Beer (1937) demonstrated, beyond reasonable doubt, that the squamosal is homologous in all tetrapods.

The squamosal has received considerable attention in the study of the mammal-like reptiles and early

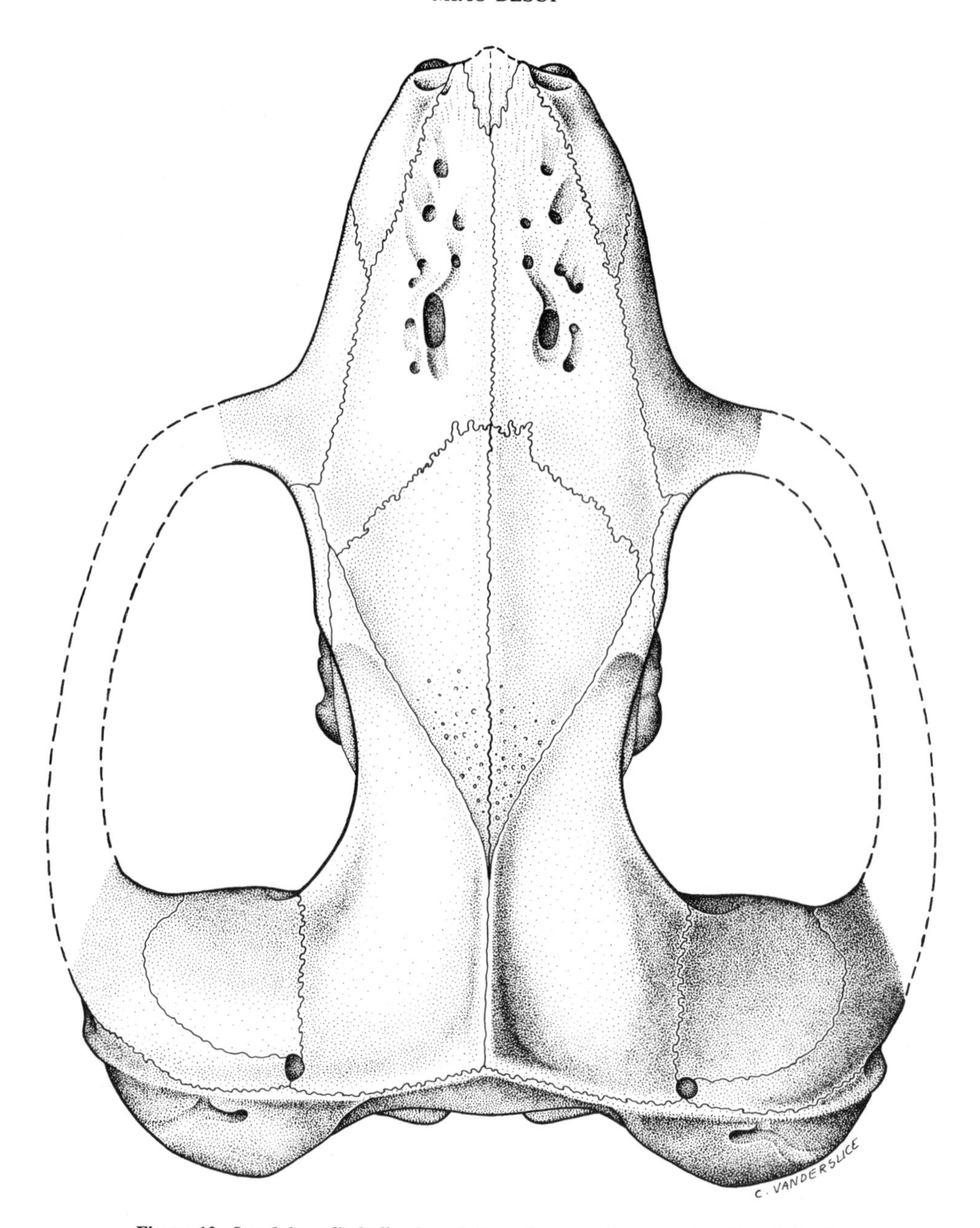

Figure 12. ***Lambdopsalis bulla*, dorsal view of composite reconstruction of skull.**

mammals, simply because the current and widely accepted arbitrary criterion of defining the class Mammalia is the presence of a squamosal-dentary joint as the major functional complex of jaw movement (Crompton and Jenkins, 1979). The best known squamosal in early mammals is that of *Morganucodon*, as described by Kermack and others (1981). As in *Lambdopsalis*, the squamosal in *Morganucodon* is preserved fairly completely, with only the anterior part of its zygomatic process missing.

Until recently, the squamosal in multituberculates was poorly known due to poor preservation. Gidley (1909) described the glenoid fossa in *Ptilodus*, without mentioning the squamosal itself. Broom (1914) pointed out that sutures among the squamosal, parietal, and other neighboring bones cannot be made out in *Ptilodus*. Nevertheless, he reconstructed an extremely large squamosal for *Taeniolabis*. However, even though Broom (*ibid.*) said it forms ". . . a large part of the cranium proper" (p. 129), the exact outline of the squamosal can be comprehended neither from his description nor from his illustration. Broom also reconstructed a large squamosal contribution to the zygomatic arch in *Taeniolabis*. In his reinvestigation, Simpson (1937) confirmed Broom's reconstruction of the zygomatic process, but did not comment on the squamosal's contribution to the "cranium proper." Simpson (*ibid.*) did, however, note that the squamoso-parietal suture is not visible in *Ptilodus* for the most part, and that ". . . it appears that the squamous part of the squamosal is small and that this element has almost no part in the cranial roof" (p. 745).

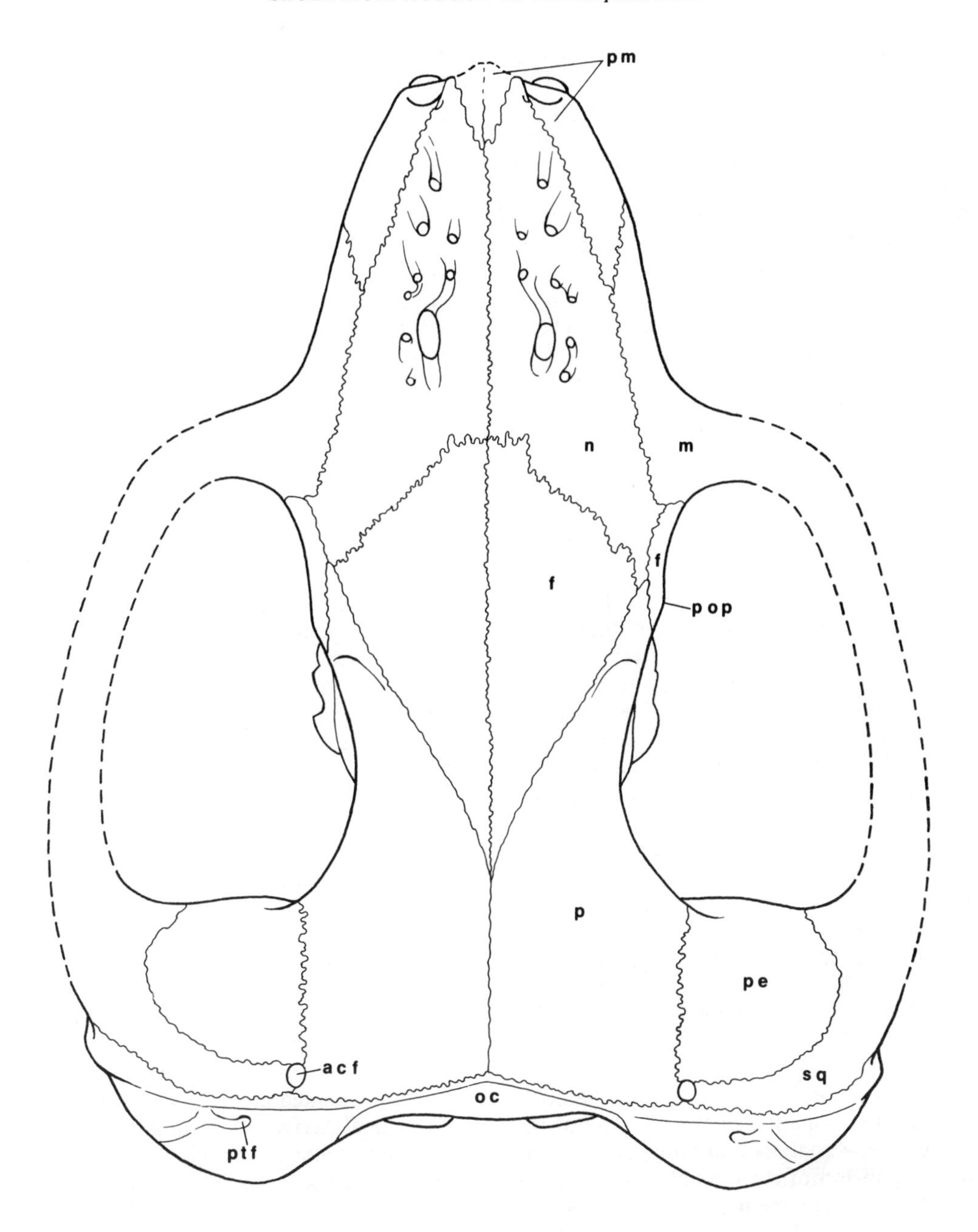

As Kielan-Jaworowska (1971) indicated, the sutures between the squamosal and its neighboring bones are not discernible in the Late Cretaceous Mongolian multituberculates. However, she and her co-workers (Kielan-Jaworowska, 1971; Kielan-Jaworowska and others, 1986) conceived of the squamosal in Mongolian taxa as an insignificant contributor to the braincase, limited to its contribution to the strong lambdoidal crest and zygomatic arch. The squamosal of *Lambdopsalis* described herein confirms that their prediction was essentially correct. Although the anterior limit of the zygomatic process of the squamosal cannot be determined in *Lambdopsalis*, it may be determined in some Mongolian multituberculates and is, in fact, well seen in *Ptilodus* (Hopson, *personal communication*).

Sloan (1979) also reconstructed a similarly small squamosal in *Ectypodus*, which is restricted in the posterior part of the skull and mainly forming the strong zygomatic process. The zygomatic process bears a relatively large glenoid fossa on its ventral surface. There is no indication of a strong cranial process of the squamosal in *Ectypodus*. Therefore, it seems reasonable to doubt the validity of Broom's reconstruction of a large squamosal contribution to the cranial wall in *Taeniolabis*. A well-preserved skull of *Taeniolabis* under study by Nancy Simmons may throw some light on this problem (Simmons, *personal communication*).

Kermack and Kielan-Jaworowska (1971) and Kermack and others (1973; 1981) noted that, unlike in therian mammals, the squamosal contributes insignificantly to the formation of the lateral wall of the braincase in morganucodontids, monotremes, and multituberculates. They also suggested this as one of the characters diagnosing their subclass Atheria (*i. e.*, notherians). Although these authors' observation is correct, the character clearly represents the plesiomorphous condition, which probably characterizes all mammals except modern therians. Therefore, it cannot be used to diagnose a monophyletic group.

Contrariwise, in arguing against a basic dichotomy (*i. e.*, nontherian versus therian) within the Mammalia as advocated by Kermack and others (*ibid.*), Kemp (1983) used expansion of the cranial process (*i. e.*, squamous part) of the squamosal to downplay the difference in braincase structure between Kermack's Atheria and the Theria. Kemp (*ibid.*, p. 374) pointed out: "It is possible that the expansion of the cranial process of the squamosal in modern therians affected the later developmental stages of the braincase, prevented the anterior lamina from fusing any longer with the periotic, and thus caused it to fuse with the alisphenoid instead."

Although Kemp's hypothesis appears at first sight to be plausible, evidence from developmental studies suggests that this may not be the case. De Beer (1937) noted that the cranial process of the squamosal in mammals develops later than its zygomatic process. Presley (1980, p. 161) also suggested: ". . . the appearance of the squamosal in the neurocranium can be seen as a consequence of the expansion of the middle and posterior cranial fossae" in modern therians. Topographically, the squamosal is lateral to both the alisphenoid and petrosal. In other words, due to the expansion of the braincase in modern therians, the alisphenoid fails to suture with the petrosal either partly or completely, and thus leaves an intervening space in the cranium. The cranial process of the squamosal fills this space in later developmental stages, and thus becomes ". . . in part a constituent of the wall of the braincase" (Presley, 1980, P. 161). It seems clear that, both in adult primitive mammals and in early developmental stages of modern therians, the squamosal does not contribute to formation of the side wall of the braincase. Phylogenetic analysis and ontogenetic study both show that expansion of the cranial process of the squamosal is a later development. Therefore, it seems unlikely that a later-emerged structure could have interfered with the earlier development of the cranial structural pattern, as Kemp has suggested.

In addition, a high degree of variation in relative size and position of the cranial process of the squamosal in modern therians seems not to change the overall similarity in their basic braincase pattern. Moreover, as I described earlier, the squamosal in *Lambdopsalis* only covers the anterolateral surface of the petrosal, and thus does not extend onto the side wall of the braincase. Yet a relatively large alisphenoid sutures with a diminished anterior lamina of the petrosal. Apparently, this also upsets Kemp's hypothesis. Furthermore, a recent study by Maier (1987) suggests that developmental patterns in the mammalian braincase may be more variable than Presley and Steel (1976) and Presley (1981) proposed. This problem will be pursued later.

To sum up: (1) lack of participation of the squamosal in forming the lateral wall of the braincase in various groups of early mammals is a plesiomorphous character, and thus cannot be used as a synapomorphy to define the so-called Atheria; (2) expansion of the cranial process of the squamosal in most, but not all, modern therians may not be a causal factor that can satisfactorily explain the presumed difference in braincase structure between Theria and so-called Atheria; and, however, (3) expansion of the cranial process of the squamosal may be considered as a synapomorphy of modern therians, closely related to expansion of the middle and posterior cranial fossae.

Further, in *Morganucodon*, Kermack and others (1981) stated that the contribution of the squamosal to the walls of the post-temporal canal is greater than that of the petrosal. They also stated that at least in two specimens of *Morganucodon*, ". . . the squamosal forms the complete posterior opening to the post-temporal canal" (Kermack and others, 1981, p. 77). Actually, in Figure 67 on p. 78, Kermack and others (1981) showed that the post-temporal canal in *Morganucodon* is completely within the squamosal. In *Lambdopsalis*, however, the reverse is the case. This is also basically true of *Catopsalis* (see Kielan-Jaworowska and others, 1986). However, whereas the post-temporal canal is walled mainly by the petrosal and insignificantly by the squamosal in *Catopsalis*, the canal is intramural within the petrosal and only its anterior opening is covered by the squamosal in *Lambdopsalis*. *Lambdopsalis* is similar to *Catopsalis* but different from *Morganucodon* in the more dorsal position of the post-temporal canal. *Lambdopsalis* is also similar to *Catopsalis* in that the post-temporal canal leads anteriorly into the ascending canal. But *Morganucodon* is more similar to cynodonts and monotremes in this aspect, in which the post-temporal canal leads anteriorly into the orbitotemporal fossa.

In summary, the post-temporal canal in multituberculates is essentially contained intramurally within the petrosal, situated more dorsally, and leads anteriorly into the cranial cavity (intracranially). In contrast, the post-temporal canal in cynodonts, morganucodontids, and monotremes is essentially bordered by the squamosal, situated more ventrally, and leads anteriorly into the orbitotemporal fossa (extracranially). Although this difference may not denote important phylogenetic significance, it is instructive to compare this difference between multituberculates and monotremes with those few presumed similarities between them which have been used to develop the argument of their relatedness.

Finally, the degree of the squamosal's contribution to the zygomatic arch is uncertain due to breakage of the zygomatic process.

PALATAL COMPLEX

Palatine (Figs. 3, 13, 14)

Description

The palatine is well preserved in the following specimens: IVPP V5429, V7151.2, V7151.5, V7151.8, V7151.52-54, and V7151.85.

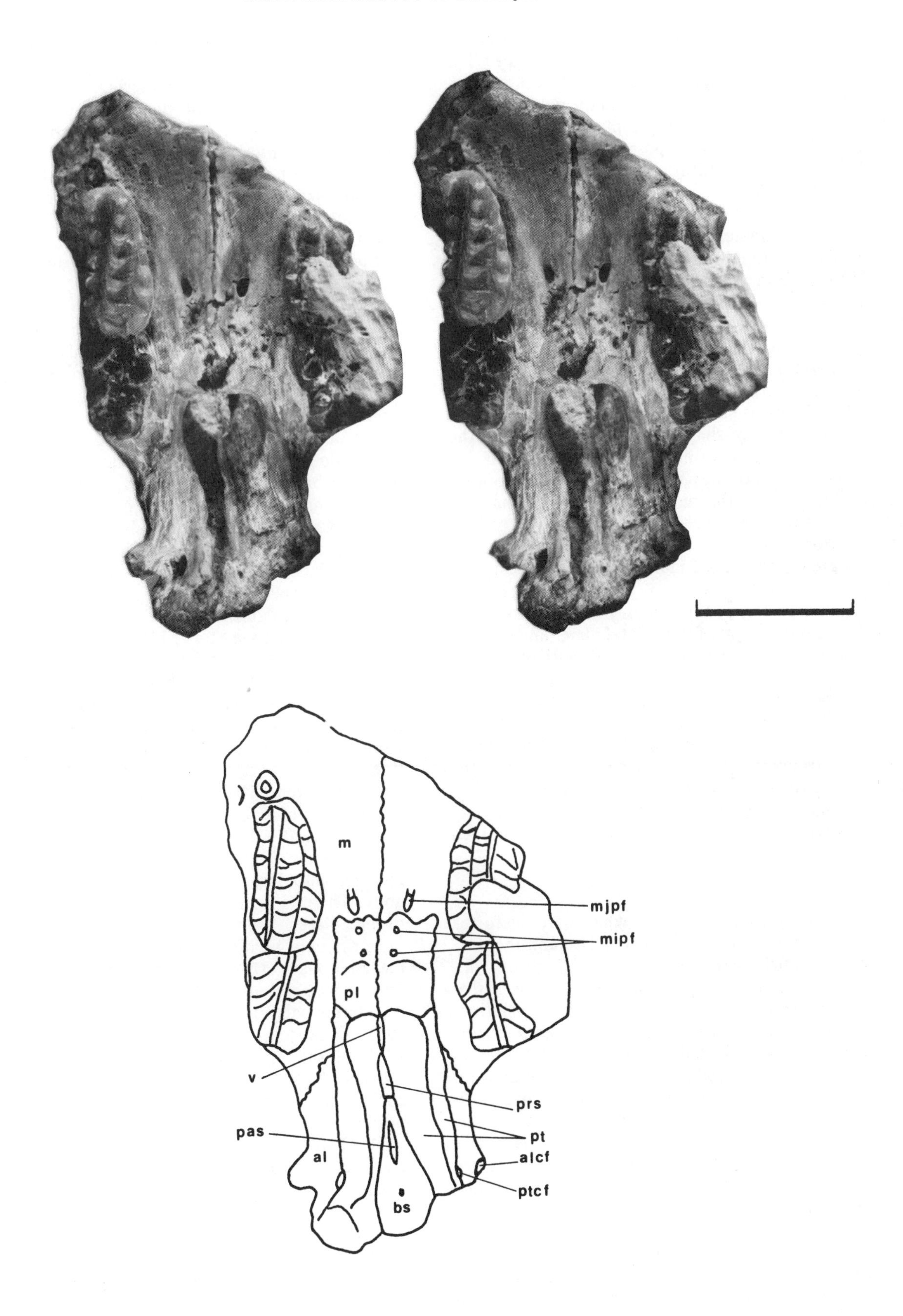

Figure 13. ***Lambdopsalis bulla*, V7151.5, stereophotograph, ventral view of fragment of skull. Anterior to top.**

The palatine consists of two parts, a palatal (or horizontal) plate and a nasal perpendicular plate. The palatine is in sutural contact with the maxilla both anteriorly and laterally, with the vomer and the palatine of the opposite side medially, with the presphenoid posteromedially, and with the pterygoid posteriorly.

The palatal plate forms the posterior part of the secondary palate. The transverse suture between the palatal plate of the palatine and the palatal process of the maxilla is approximately at the level of the posterior one-third of M1. The longitudinal suture between the palatal plate of the palatine and the alveolar process of the maxilla is parallel to the tooth row and extends posterolaterally. At about the level of the last two cusps of the internal row of M1, there are two foramina in the palatal plate of the palatine, with the anterior one larger than the posterior one (here designated as minor palatine foramina). The choana is nearly as broad as the posterior part of the palatal plate of the palatine. In ventral view, the rim of the choanae is formed by the posterior margins of the palatal plates of the palatines, and situated approximately at the level of the middle cusp of the internal row of M2. Starting at the level of the anterior third of M2, the posterior part of the palatal plate of the palatine becomes greatly thickened into a torus (*i. e.*, postpalatine torus).

The nasal or perpendicular plate of the palatine arises from the lateral margin of the palatal plate, and extends dorsally before turning medially and horizontally to suture with its counterpart on the other side above the choanal opening. Thus, the palatine forms the dorsal, lateral and ventral walls of the choana. The nasal plate of the palatine is overlapped laterally mostly by the maxilla, and partially (only in its most posterior part) by the alisphenoid. In ventral view, the most posterior part of the nasal plate wedges between the alisphenoid laterally and the pterygoid medially. The palatine is excluded entirely from the orbit.

Discussion

The palatine is a dermal bone, which partially forms the posterior part of the secondary palate and the dorsal, ventral, and lateral walls of the choana. Together with the pterygoids and the vomer (the latter has been discussed in connection with the nasal cavity), the palatines constitute most of the palatal complex in *Lambdopsalis.*

Minor palatine foramina.—Two small foramina posterior to the major palatine foramen are present on either side of the palatal plate of the palatine in *Lambdopsalis*, here designated as minor palatine foramina. Although their number and exact position(s) may vary, similar foramina (or a foramen) also occur(s) in *Kamptobaatar* (see Kielan-Jaworowska, 1971), *Chulsanbaatar*, *Nemegtbaatar* (see Kielan-Jaworowska and others, 1986), and *Ptilodus* (see Simpson, 1937) among multituberculates. The foramina (or a foramen) occupying similar topographical positions also occur(s) in many modern mammalian groups including humans. It also has been reported in *Morganucodon* (see Kermack and others, 1981), though with different terminology (their "lesser palatine foramen"), and probably is not homologous (see discussion below).

Simpson (1937) first described a few small and irregular foramina in the posterior part of the palatal plate of the palatine, just anterior to the postpalatine torus, in *Ptilodus*. He did not name them, nor did he speculate upon their possible function. Although Kielan-Jaworowska (1971) did term them as minor palatine foramina in *Kamptobaatar*, she did not postulate a function for them. Kielan-Jaworowska and others (1986) applied the term to the similar foramen (or foramina) in *Chulsanbaatar* and *Nemegtbaatar*.

As discussed earlier, the function of the major palatine foramen has been interpreted with reasonable confidence based upon modern mammals. By the same token, function of the minor palatine foramina in *Lambdopsalis* as well as in the other multituberculate taxa can be interpreted with the same degree of confidence as that based upon function in their modern homologues. In many modern mammals, the sphenopalatine ganglion gives off one or several branches, called minor palatine nerves, which pass through the minor palatine foramina to innervate mainly the soft palate as well as the gum. Accompanying the minor palatine nerves through the minor palatine foramina are minor palatine arteries (*i. e.*, the branches arising from the descending palatine artery); they are nutrient to the minor palatine nerves and the palate (*e. g.*, see Story, 1951, for Procyonidae; and Reighard and Jennings, 1935, for *Felis domesticus*). The same is true of humans although terminology is different (see Romanes, 1981). In human anatomy, sometimes the term "pterygopalatine ganglion" replaces the term "sphenopalatine ganglion," and the term "lesser palatine nerve" or artery replaces the term "minor palatine nerve" or artery. Therefore, based upon the fact that the minor palatine foramina (or foramen) occur(s) in phylogenetically diverse groups within Mammalia with similar topographic position(s) and similar function, the most parsimonious interpretation is that they are homologous. There is no reason to suggest otherwise in multituberculates.

It should be noted, however, that topographically the lesser palatine foramen in *Morganucodon* as described by Kermack and others (1981) is more reminiscent of the "posterior palatine notch" in *Procyon lotor* (see Story, 1951), or the "posterior maxillary notch" (see Wahlert, 1974) or the "palatine vein foramen" (Hill, 1935) in rodents. As Wahlert (1974) noted, the notch is actually enclosed as a foramen in many rodents. It transmits the posterior palatine artery and large palatine vein (Story, 1951). I strongly suspect that the "lesser palatine foramen" as identified by Kermack and others (1981) in *Morganucodon* is too large and too posterolaterally placed to qualify for that designation. Thus it should not be equated with the minor palatine foramen (or foramina) in other mammals.

Exclusion of palatine from orbit.—The palatine is excluded entirely from the ventral part of the orbit in *Lambdopsalis*. Simpson (1937, p. 746) noted: "The palatine seems to have little, if any, exposure. . ." in the orbital wall of *Ptilodus*. Due to the usually poor preservation in this region, the orbital exposure of the palatine in most other known multituberculate cranial materials is not certain. Although Kielan-Jaworowska (1971) and Sloan (1979) reconstructed a significant orbital exposure of the palatine in *Kamptobaatar* and *Ectypodus* respectively, the sutural delineations of the individual bone within the orbital mosaic are not visible in either case. Thus the presumed large orbital exposure of the palatine in those two genera is at best conjectural.

At present, nevertheless, we should base our phylogenetic interpretation of this character in multituberculates upon the positive evidence about the exclusion of the palatine from the orbit in *Lambdopsalis* and *Ptilodus*. Different contribution of various bones to the orbital mosaic has been emphasized in phylogenetic analyses of early eutherians (*e. g.*, Novacek, 1980; 1986). Following Muller (1935), Novacek (1980; 1986) considered diminution of the orbital exposure of the palatine in eutherians as a derived character. Although Muller (1935), in the same paper, was self-contradicted by stating: "From the shifting of the nasal capsule we might conclude the following primitive character: the maxilla is located in the oral basal part of the orbit whence the palatine has disappeared" (p. 136), the author did point out that the palatine being ". . . ousted from the oral part of the orbit" (p. 137) is a secondary character. Muller (*ibid.*) also suggested that this may be due to the extension of the maxilla in the posterior direction. Simpson (1937) also noted that the exclusion of the palatine from the orbit in *Ptilodus* ". . . may be correlated with the unusually heavy alveolar and zygomatic development of the maxilla" (p. 746). It should be added that the posterior extension of the alveolar maxilla (which laterally overlaps the palatine) in *Lambdopsalis* and *Ectypodus* (and possibly in *Ptilodus* as well) causes a direct contact between the maxilla and the alisphenoid. Exclusion of the palatine from the orbit in *Lambdopsalis*, in turn, results in a wide contact between the orbital process of the frontal and the ascending process of the maxilla.

In summary, considering that the orbital exposure of the palatine occurs in many cynodonts (*e. g.*, see Crompton, 1958 for *Diarthrognathus*; Kemp, 1979 for *Procynosuchus*, Kemp, 1980 for *Luangwa*; and Sues, 1986 for *Kayentatherium*) and possibly in *Morganucodon* (see Kermack and others, 1981), the exclusion of the palatine from the orbit might be a synapomorphy for multituberculates.

Pterygoid (Figs. 3, 13-15, 18)

Description

The pterygoid is best preserved in IVPP V5429, V7151.5, V7151.51, V7151.54, V7151.81, V7151.91, and V7151.92; and less well preserved (but also provided valuable information) in V7151.2, V7151.8, V7151.52, V7151.53, V7151.79, and V7151.90.

The pterygoid consists of a horizontal basal plate, and a vertical lamina (usually referred to as the lateral flange or ridge of the pterygoid). The basal plate of the pterygoid is in contact dorsally with the ventral surfaces of the presphenoid and basisphenoid. The basal plate, however, does not meet its counterpart on the other side in the midline, thus leaving the ventral surface of the presphenoid and basisphenoid partly exposed. The lateral flange of the pterygoid descends from the basal plate to form the lateral wall of the nasopharyngeal duct.

The pterygoid is in sutural contact with the palatine anteriorly, with the alisphenoid laterally, and with the presphenoid and basisphenoid dorsomedially. In both V5429 and V7151.81, the posterior edge of the lateral flange of the pterygoid extends posteriorly a little beyond (and ventral to) the anterior tip of the promontorium of the petrosal.

There is an anterodorsally/posteroventrally oriented bony canal between the alisphenoid and the lateral flange of the pterygoid which is here designated as the pterygoid (or Vidian) canal. The anterior opening of the canal lies ventral to the sphenorbital fissure and ventromedial to the anterior opening of the alisphenoid canal (equivalent to "canal of ?maxillary artery" of Kielan-Jaworowska and others, 1986).

The nasopharyngeal region is dorsoventrally deep in ventral view. Except for the posterior extension of the vomer (terminated at about a level of the posterior edge of M2) in the midline and a short and low keel in the midline of the basisphenoid, there are no parasagittal ridges like those previously designated as "pterygoid" in some Late Cretaceous Mongolian multituberculates (Kielan-Jaworowska, 1970, 1971) and *Morganucodon* (Kermack and others, 1981) as well as "pterygo-palatine crest" in certain tritylodonts (Sues, 1986).

The sutures are clearly visible between the pterygoid and palatine, and between the pterygoid and presphenoid plus basisphenoid in V7151.92. However, the suture between the posterior part of the lateral flange of the pterygoid and the alisphenoid is less certain.

A well-defined ectopterygoid fossa is present between the pterygoid process of the alisphenoid (= "the ectopterygoid crest of the alisphenoid", *sensu* Novacek, 1986) and the posterior part of the lateral flange of the pterygoid (= "the entopterygoid crest of the pterygoid," *sensu* Novacek, *ibid.*).

Discussion

The multituberculate cranial materials available to Gidley (1909), Broom (1914), Simpson (1937), and Hahn (1969) lack either good preservation or sutural demarcation in the posterior part of the palatal region, and thus have not allowed identification of a pterygoid. Kielan-Jaworowska (1970; 1971) first reported the presence of "pterygoid" in *Kamptobaatar*, identified as a ridge between the lateral wall of the choana and the vomer. Consequently, each side of the choana as separated by the vomer is divided further by the "pterygoid" into two longitudinal channels. Kielan-Jaworowska (*ibid.*) considered it the most peculiar feature of the region. She went further to interpret the lateral channel as an area for muscle attachment and the medial channel as the air passage. This was essentially followed by Kermack and others

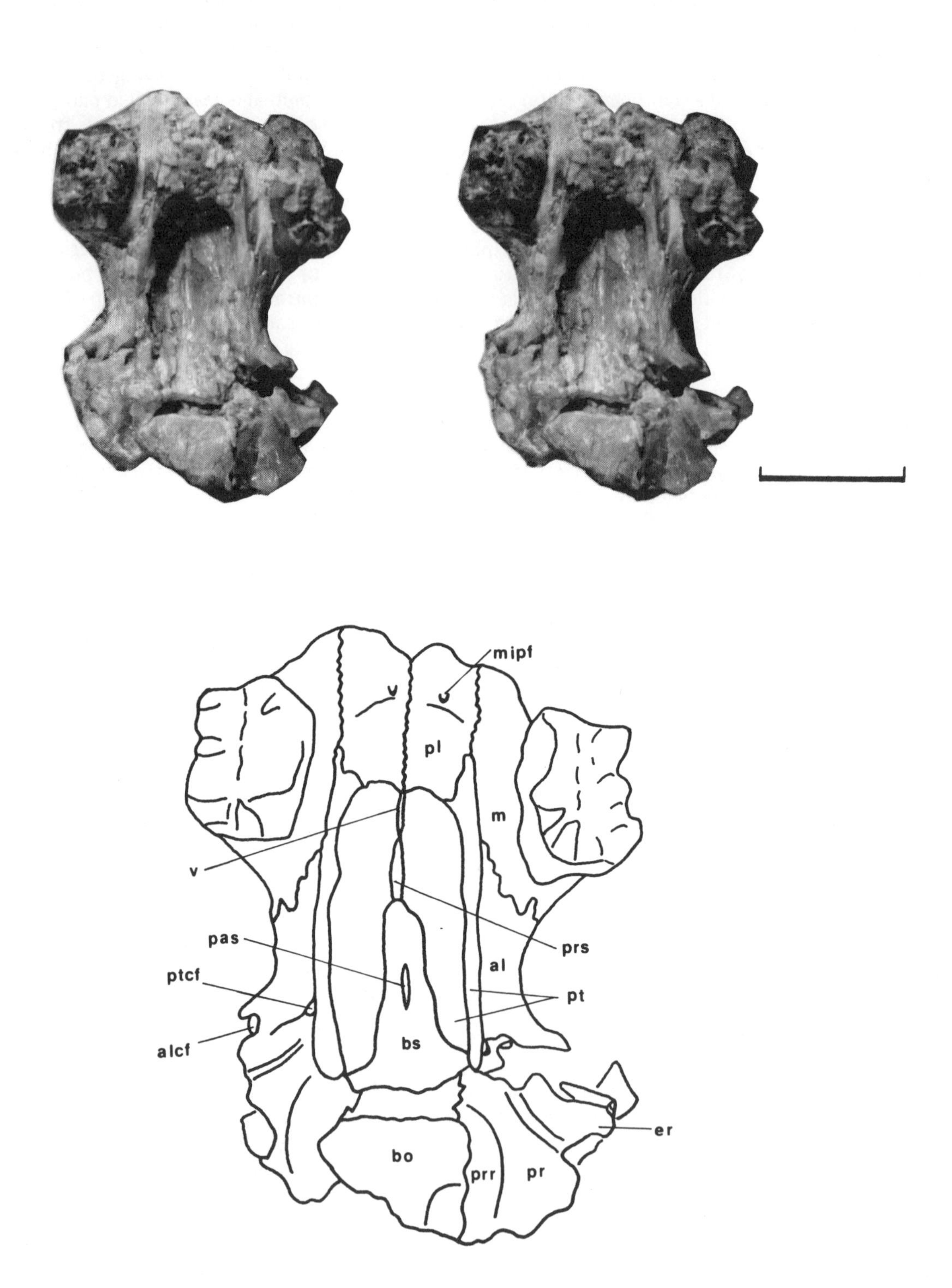

Figure 14. ***Lambdopsalis bulla*, V7151.92, stereophotograph, ventral view of fragment of skull. Anterior to top.**

(1981) for *Morganucodon*. Kielan-Jaworowska and others (1986) documented an exactly similar condition in *Nemegtbaatar*.

However, as one of Kielan-Jaworowska's co-authors, Presley (*personal communication*) did not believe that the so-called "pterygoid" in *Kamptobaatar* is indeed a homologue of the mammalian pterygoid. Instead, he regarded the "pterygoid" in *Kamptobaatar* as a ridge preventing occlusion of the auditory tube by the soft palate in a relatively shallow choana. A similar function is carried out by cartilage or elastic tissue in modern mammals. Nevertheless, he does not exclude the possibility that the ridge also served for attachment of the levator palati muscle (*personal communication*). This is consistent with (but independent from) Barghusen's (1986, p. 260) interpretation: "In *Kamptobaatar*, the lateral nasopharyngeal channel leaves the osseous confines of the nasopharynx in the position where the entrance to the auditory tube is found in mammals." It is interesting to note that Barghusen (*ibid.*) based his interpretation on the assumption of multituberculates' having a typical mammalian middle ear as in monotremes and therians, and thus believed that multituberculates already developed ". . . an interconnection between the nasopharynx and the auditory structures" (p. 259). The discovery of three ear ossicles in *Lambdopsalis* (see Miao and Lillegraven, 1986) strengthens his argument. As an aside, Barghusen (1986) called the "pterygoid" in *Kamptobaatar* the pterygopalatine ridge.

The absence of a "pterygopalatine ridge" coupled with the acquisition of a deep choana in *Lambdopsalis* provides strong support for the interpretations offered by Barghusen and Presley. In addition, Sues (1986) extended Barghusen's interpretation to explain an essentially similar pterygopalatine ridge in the tritylodont *Kayentatherium wellesi*. As Allin (1975) suggested, advanced cynodonts already may have developed a middle ear configuration characteristic of modern mammals, even though the transformation of their reptilian jaw elements into mammalian ear ossicles had not yet been accomplished. Hence Sues' extrapolation seems reasonable, and I have no qualms about the explanatory power of Barghusen's and Presley's interpretation as applied to the "pterygoid" of *Morganucodon*. Furthermore, the clear sutural delineation of the pterygoid in *Lambdopsalis* shows that its shape and position are characteristic of modern therians. This would imply that the search for the pterygoid in other multituberculates should be directed to a similar, conventional position.

Homology of mammalian pterygoid and problem of ectopterygoid.—Homology of the mammalian pterygoid has generated considerable discussion in the literature of both developmental biology and paleontology (*e. g.*, De Beer, 1929, 1937; Parrington and Westoll, 1940; Presley and Steel, 1978; and Moore, 1981). Although opinions vary among various authors, all agreed that two elements, a dorsal and a ventral blastema, were involved in forming the pterygoid, recognizable during the ontogenic development of marsupials and eutherians. Thus, the compound structure of the pterygoid in modern therian mammals is beyond reasonable doubt. The traditional view has been that the dorsal element represents the detached lateral part of the parasphenoid, and the ventral element is homologous with the reptilian pterygoid (*e. g.*, De Beer, 1929, 1937). Parrington and Westoll (1940), however, argued forcefully that the dorsal element was derived from the pterygoid of therapsids, and the ventral element from the ectopterygoid, based mainly upon paleontological evidence. Parrington and Westoll's notion was strongly supported more recently by Presley and Steel's (1978) developmental studies on a variety of monotremes, marsupials, and eutherians. Still another view represented by Watson (1916) and Kesteven (1918) held that the dorsal element was derived from the reptilian pterygoid and the ventral element from the alisphenoid. De Beer and Fell (1936) showed that, at embryonic "stage 5" (122 mm) of *Ornithorhynchus*, the alisphenoid is present, lying lateral to the dorsal element and anterodorsal to the ventral element, and therefore cannot be homologous with the ventral element. Presley and Steel (1978) further pointed out that the ventral part, *i. e.*, hamulus, is derived from an ancestral membrane bone (*i. e.*, ectopterygoid) and becomes a secondary cartilage in many ditrematous mammals (a suggestion made earlier by Gaupp [1901] and De Beer [1929]). They also noted that the processus pterygoideus or "pterygoid cartilage" of the ala temporalis (*i. e.*, alisphenoid) lies lateral to the dorsal element. Therefore, they strongly rejected the possibility of homology between the pterygoid process of the alisphenoid and the ventral element of the mammalian composite pterygoid.

However, Presley and Steel (1978) suggested that in monotremes the association between the processus pterygoideus (or "pterygoid cartilage") of ala temporalis and the pterygoid ". . . is very close and the ossification spreads from the pterygoid into the subjacent cartilage of the ala temporalis. Comparison with the cynodont *Thrinaxodon* shows that membrane bone in this position could very well be derived from that portion of the pterygoid which lies above the pharynx and includes the pharyngeal ridges of the fossil" (p. 106).

It is interesting to note that, from Presley and Steel's plausible suggestion, Sues (1986) derived a confusing statement (or rather a misrepresentation of Presley and Steel) about ". . . Presley and Steel's (1978: 106) suggestion that the lateral flange of the pterygoid is homologous to the hamulus pterygoidei of the mammalian ala temporalis" (p. 234). In other words, Presley and Steel suspected that part of the ossification in the pterygoid process of the alisphenoid could be derived from the lateral flange of the pterygoid. However, it does not imply that the *whole* lateral flange of the pterygoid is homologous to the pterygoid process of the alisphenoid (= "the hamulus pterygoidei of ala temporalis" of Sues) as Sues interpreted. In addition, the hamulus is usually referred to the ventral element of the pterygoid in mammals, and thus Sues' usage of "hamulus pterygoidei" of ala temporalis itself is confusing.

Furthermore, Presley and Steel (1978) argued that the inferred homology of the hamulus with the reptilian ectopterygoid rather than with the secondarily chondrified pterygoid process of the ala temporalis is also supported by the coexistence of the hamulus and the pterygoid cartilage of the ala temporalis. More recently, Kohncke (1985) documented three different conditions of the pterygoid process in mammalian carnivores: (1) in *Cryptoprocta*, the pterygoid process of the ala temporalis remains completely separated both from the pterygoid and from the hamulus; (2) in *Felis*, the pterygoid process of the ala temporalis is in synostosis with the pterygoid, with the pterygoid (or "Vidian") canal marking the old suture; and (3) in *Paradoxurus*, the pterygoid process of the ala temporalis and the cartilaginous hamulus are continuous. As an aside, the last condition is also seen in *Sorex* (see De Beer, 1929).

From the above review, therefore, it is reasonably certain to assume homology of the mammalian hamulus with the reptilian ectopterygoid. This assessment accords well with paleontological evidence. Since Parrington and Westoll's (1940) review, much more has been learned about cranial anatomy of advanced cynodonts and early mammals. Had the information been available to these two authors, they could have made an even stronger case. For example, we now know that the ectopterygoid is not recognizable as a separate bone in *Diarthrognathus* (see Crompton, 1958), *Oligokyphus* (see Kühne, 1956), *Bienotherium* (see Young, 1947; Hopson, 1964), *Kayentatherium* (see Sues, 1986), *Tritylodon* (see Gow, 1985), and *Probainognathus* (see Romer, 1970). Gow (1986) did not recognize an ectopterygoid in *Megazostrodon*. The definite absence of an ectopterygoid in *Lambdopsalis* gives added support for the interpretations of Parrington and Westoll (1940) and Presley and Steel (1978), which show that, unlike most reptiles, the ectopterygoid in advanced cynodonts and early mammals was already incorporated into the pterygoid as a ventral element. It should be noted that the reported presence of ectopterygoid in *Morganucodon* and many multituberculates was based upon dubious extrapolations from the presence of an ectopterygoid in modern monotremes, and cannot be borne out from actual observations on the fossils (*personal observations*). The morphology of modern monotremes represents a melange of retained primitive features since their early divergence from the mainstream of mammalian evolution and many specializations owing to adaptations to their specialized lives. Hence the morphology of monotremes necessarily neither represents the structural link between reptiles and mammals nor serves as a stereotype for the radiation of early mammals. In this particular case, we should not let understanding of the posterior palatal structure in early mammals be constrained by retention of a separate ectopterygoid or "echidna pterygoid" in monotremes.

Finally, it is clear, at least in *Lambdopsalis* among multituberculates, that a ditrematous mammalian condition of the palatal structure has been essentially achieved. Even though not prominent, a ventroposteriorly projected process of the lateral flange of the pterygoid in *Lambdopsalis* could well be interpreted as the hamulus.

Pterygoid canal.—The pterygoid (or "Vidian") canal in therians is a bony canal either ". . . passing along or through site of fusion of pterygoid bone and basisphenoid" (MacPhee, 1981, p. 61), or passing along or through the junction of pterygoid and alisphenoid (Terry, 1917; Kielan-Jaworowska and others, 1986). Although this structure has been described in some previous paleontological literature dealing with cranial anatomy both of cynodonts and early mammals (*e. g.*, Parrington, 1946; Kemp, 1980; Kermack and others, 1981; and Sues, 1986), Kielan-Jaworowska and others (1986) called attention to the possible misnomer, and I could not agree more. For instance, Kemp (1980, Fig. 9 and p. 214) showed a pair of foramina anterolateral to the pituitary fossa (and facing posteriorly) as the posterior openings for the Vidian canals in *Luangwa*. This implies that the Vidian nerve has at least a partially endocranial course, which is quite different from its partially intramural but completely extracranial course as commonly seen in mammals. Sues (1986) attempted to homologize the Vidian canal of *Luangwa* with his interpterygoid vacuity in *Kayentatherium*, and thought that the latter also would "mark the reentry into the skull of the ramus palatinus of N. facialis" (p. 236). Apparently, Sues also implied that the nerve has a partially endocranial course. However, he failed to point out where the anterior exit of the nerve is and where the nerve leads after its entry into the interpterygoid vacuity in *Kayentatherium*. Kermack and others (1981) described a Vidian nerve groove rather than a canal along the base of the palatine in *Morganucodon*. Kielan-Jaworowska and others (1986) realized the problem, but for convenience they still called a groove associated with the "hiatus Fallopii" as "Vidian" in the multituberculates by considering its possible functional correlation. Therefore, probably in none of the above-discussed cases, does the so-called "Vidian canal" conform to the definition of the term as applied to modern mammals. This problem will be further discussed in the section entitled "Petrosal and Auditory Region."

As described earlier in *Lambdopsalis*, a bony canal extends anterodorsally/posteroventrally between the alisphenoid and the lateral flange of the pterygoid, and therein was designated as the pterygoid (or "Vidian") canal. Its anterior opening lies both ventral to the sphenorbital fissure and ventromedial to the anterior opening of the alisphenoid canal, and also marks the posterior termination of the sphenopalatine groove. The posterior opening of the pterygoid canal is ventral to the posterior opening of the alisphenoid canal. A distinct groove, here designated as the Vidian groove, descends anteroventrally from an opening designated as the hiatus Fallopii (for further discussion see section "Petrosal and Auditory Region") anteroventral to the foramen ovale inferium, and extends to the posterior opening of pterygoid canal. [However, this groove is not equivalent to "?vidian groove" of Kielan-Jaworowska and others, 1986]. This mutual topographic relationship between the Vidian groove and the pterygoid canal strengthens the designation of the canal as well as that of the groove. If these designations are correct, *Lambdopsalis* becomes the first known fossil "nontherian" mammal to have a pterygoid canal that completely fits the definition of the structure as applied to modern therian mammals (*i. e.*, an intramural bony canal lying between the pterygoid and alisphenoid).

Particularly, *Lambdopsalis* appears similar to *Felis domesticus* in that the pterygoid canal marks the sutural zone between the pterygoid and alisphenoid, whereas the suture itself disappeared in adult skulls. Based upon his classic study of the primordial cranium of the cat, Terry (1917, p. 381) pointed out: ". . . the Vidian nerve runs along the mesenchymal junction of the pterygoid cartilage and pterygoid process of the ala temporalis; now, although no suture or line can be found in the adult skull indicative of the original limit of the pterygoid bone toward the alisphenoid, . . . yet the course of the bony walled Vidian canal of the adult can be taken as marking this boundary."

The pterygoid canal in modern therians transmits the Vidian nerve, which is formed by the greater superficial petrosal nerve (palatine branch of the facial nerve) and the deep petrosal nerve (a branch of the internal carotid plexus). The Vidian nerve runs forward to join the sphenopalatine ganglion. Occasionally, an accompanying Vidian artery also passes through the canal (see also MacPhee, 1981; Kielan-Jaworowska and others, 1986). However, MacPhee (1981, p. 61) cautioned: "Mere presence of identifiable pterygoid canal in a dried skull or fossil does not conclusively establish that artery of pterygoid canal was present, in absence of other evidence. There is no way of determining in such cases whether canal contained only nerves or the artery as well." Accordingly, Kielan-Jaworowska and others (1986) reconstructed a Vidian artery in some Late Cretaceous Mongolian multituberculates as accompanying the greater superficial petrosal nerve, ". . . passing close to the internal carotid artery, and then on the lateral wall of the choana" (p. 563). Because there is no real pterygoid canal described in those Mongolian taxa, their reconstruction of the route for the Vidian artery cannot be reasonably extrapolated to *Lambdopsalis*. On the other hand, due to its highly variable connections as observed in modern mammals (MacPhee, 1981), it is impossible to ascertain if the Vidian artery passed into the pterygoid canal in *Lambdopsalis*.

Alisphenoid canal.—Although the alisphenoid canal grooves the medial side of the alisphenoid in *Lambdopsalis* and perhaps should be treated in the section entitled "Alisphenoid," it seems more convenient to discuss it in the present context.

The supposed presence both of an alisphenoid canal and an alisphenoid bulla was reported in *Ptilodus* by Gidley (1909). Whereas Broom (1914) doubted the presence of the alisphenoid bulla but did not comment on the presence or absence of the alisphenoid canal, in his restudy of Gidley's specimen of *Ptilodus* Simpson (1937) suggested that probably neither the alisphenoid bulla nor the alisphenoid canal was present in *Ptilodus*. More recently, Hahn (1981) described a rather long "channel" in *Pseudobolodon* extending from the "foramen pseudovale inferium" anteriorly to the "sphenorbital canal," and regarded it as containing a blood vessel in life. Kielan-Jaworowska and others (1986) suggested that the channel in *Pseudobolodon* might correspond to the canal of the maxillary artery as designated in their description of ?*Catopsalis joyneri* and unidentified Hell Creek petrosals. They also attempted to equate the canal of the maxillary artery with Watson's (1916) "alisphenoid canal" in *Ornithorhynchus*. However, although the above-referred canal or channel may contain the blood vessel of the same name (*i. e.*, maxillary or internal maxillary artery), the canal of the maxillary artery as described by Kielan-Jaworowska and others (1986) is the ". . . canal from facial sulcus to foramen near foramina masticatorium and ovale inferium" (p. 589). Moreover, Watson's "alisphenoid canal" in the platypus and Hahn's "channel" in *Pseudobolodon* run from the anterior margin of the foramen ovale or the foramen ovale inferium *anteriorly* to the sphenorbital fissure. In other words, whereas Kielan-Jaworowska and others' canal of the maxillary artery contains the posterior section of the artery in ?*Catopsalis joyneri* and the unidentified Hell Creek petrosals, Watson's alisphenoid canal in the platypus and Hahn's "channel" in *Pseudobolodon* contain the more anterior section of the artery. Apparently, they described bony enclosures of different parts of the same artery.

Both canals can be observed in *Lambdopsalis*, one canal posterior to the foramen ovale inferium (= canal of maxillary artery *sensu* Kielan-Jaworowska and others, 1986), and the other anterior to the foramen ovale inferium (= alisphenoid canal *sensu* Watson, 1916, as well as its general usage among many modern mammals). Therefore, I restrict the term "alisphenoid canal" in *Lambdopsalis* to refer to the canal grooving the medial side of the alisphenoid bone and extending anteriorly from the anterior margin of the foramen ovale inferium to the opening (*i. e.*, the anterior opening of the canal) lateral to the sphenorbital fissure. This usage accords well with the general term of alisphenoid canal in mammalian anatomy. I thus adopt "canal of maxillary artery" of Kielan-Jaworowska and others (1986) to name the canal posterior to the foramen ovale inferium in *Lambdopsalis*. Wible (*personal communication*), however, suggests that canal of ramus inferior of stapedial artery is a more appropriate term for this structure.

Recently, Wahlert (1985, p. 322) stated: "Many mammals lack an alisphenoid canal, and the maxillary artery runs lateral to the bone; this condition appears to be primitive, and the alisphenoid canals in different mammalian orders are not homologous." However, an equally parsimonious alternative interpretation might be as such: (1) although original absence of an alisphenoid canal in earliest mammals represents a primitive condition, the lack of the canal in many eutherians may be simply due to the secondary loss; and (2) to the best of my knowledge, when an alisphenoid canal is present, it almost invariably transmits the internal maxillary artery, which may be argued for their homologous nature. Against this alternative, one may turn the argument around: (1) due to the high degree of diversity, it is

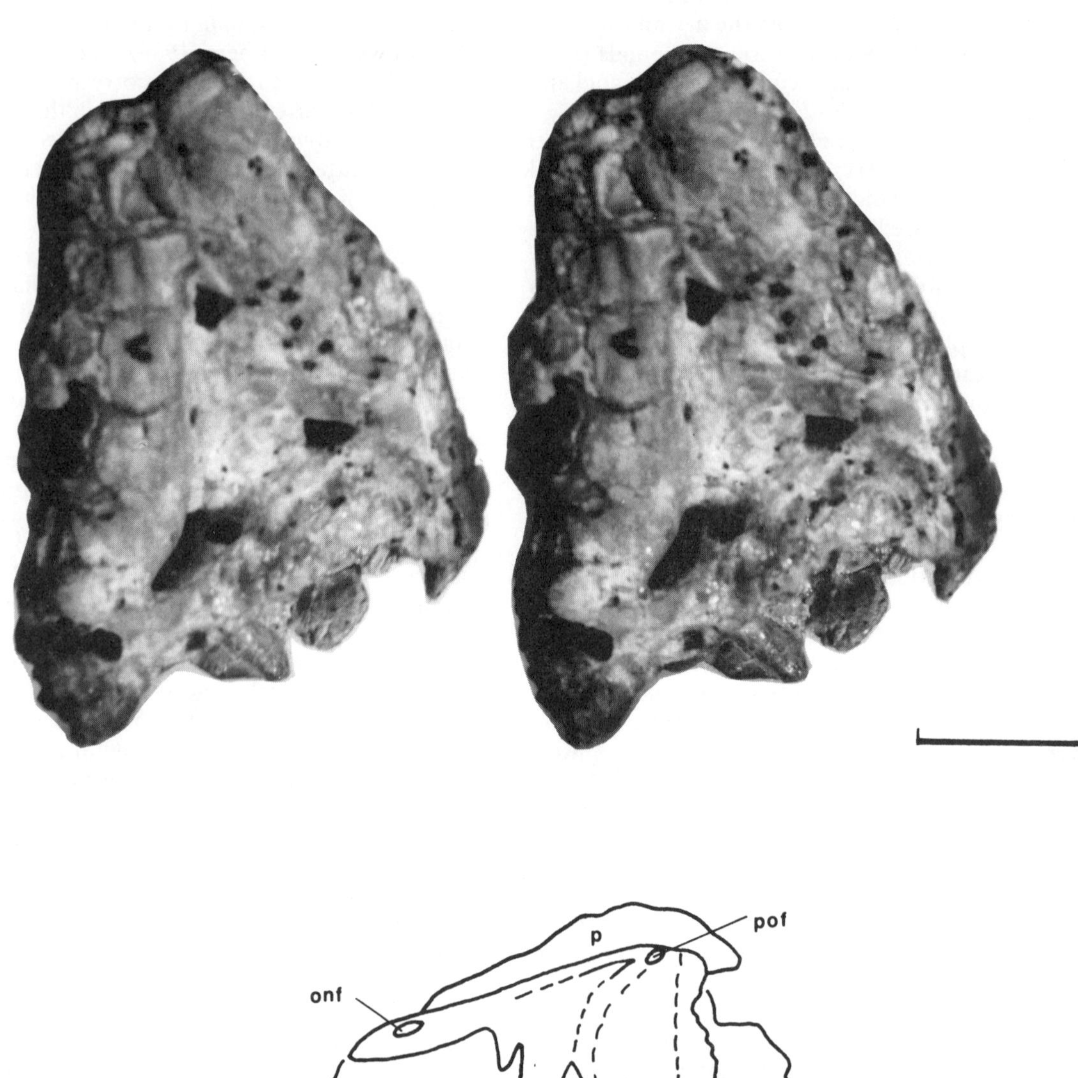

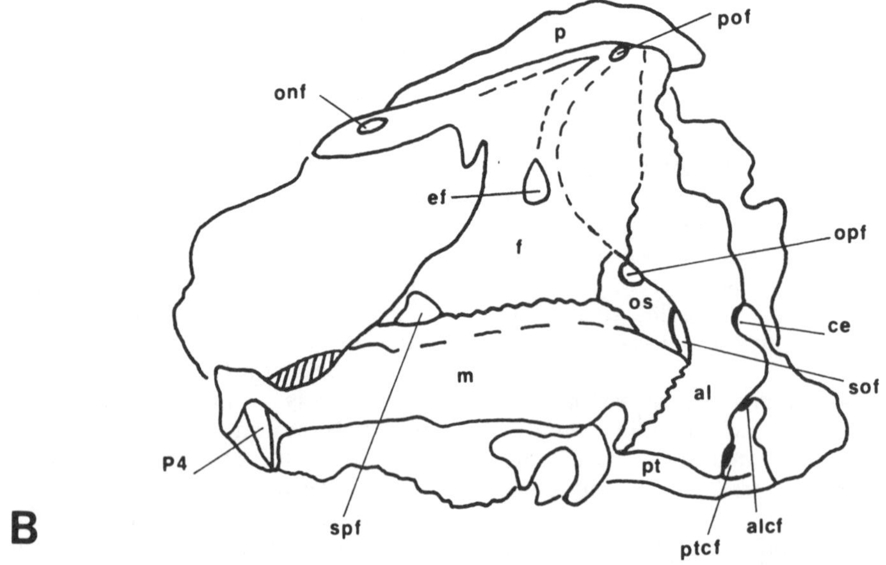

Figure 15. ***Lambdopsalis bulla*, V7151.51, lateral view of fragment of skull. *A*, stereophotograph, anterior to top; and *B*, explanatory drawing, anterior to left.**

difficult to determine the polarity of the character state concerned with presence versus absence of the alisphenoid canal. For example, even the same author sees a long alisphenoid canal as primitive and a shortened one as derived for a variety of selected eutherian orders at one time (Novacek, 1985), and regards an indistinct or absent alisphenoid canal as a primitive condition at a later time (Novacek, 1986); and (2) the internal maxillary artery is supplied via external carotid artery in some mammals, but via stapedial artery (which is originated from internal carotid artery) in the others (McDowell, 1958; Kielan-Jaworowska and others, 1986). Thus, it may imply that the internal maxillary arteries are not homologous in different mammalian groups, nor would be their partial enclosures—the alisphenoid canals. However, in the latter case, the branch of the stapedial artery is often annexed by the external carotid artery (Cartmill and MacPhee, 1980).

Perhaps it is advisable not to attach too much weight to the phylogenetic significance of a character of dubious homology such as the alisphenoid canal. Again this illustrates that, whatever school of systematics one chooses to follow, the basic problem still centers in what kind of characters one decides to employ. Furthermore, the answer to the question whether the alisphenoid canal in *Lambdopsalis* is homologous to that in some eutherians depends upon one's choice of definitions of homology (Ghiselin, 1976). However, considering that no alisphenoid canal perforates the alisphenoid in marsupials (Gregory, 1910; Archer, 1976), it seems more parsimonious to assume that the alisphenoid canal in *Lambdopsalis* evolved independently, and thus is not homologous to that in eutherians, even if the alisphenoid canals in different eutherian orders may be homologous.

BRAINCASE

Presphenoid—Basisphenoid—Parasphenoid Complex (Figs. 3, 13, 14, 18)

Description

The presphenoid and basisphenoid are well preserved in the following specimens: IVPP V5429, V5429.15, V7151.2, V7151.5, V7151.8, V7151.51, V7151.52, V7151.54, V7151.79, and V7151.91-93. The parasphenoid is recognized in V5429.15, V7151.5, V7151.54, and V7151.91.

Due to lack of clear sutural demarcation between presphenoid, basisphenoid, and parasphenoid, they are treated together in a single section for convenience.

The ventral exposure of presphenoid-basisphenoid is triangular, or wedge-shaped, with a long apex extended anteriorly to overlie the vomer. At least the posterior two-thirds of the total anteroposterior length of the presphenoid-basisphenoid actually are underlaid ventrolaterally by the basal plate of the pterygoid. There is a suture, at the level of the anterior tip of the promontorium, clearly separating the basisphenoid from the basioccipital.

In ventral view, the ill-delineated presphenoid forms a low median ridge in the anterior part of the nasopharyngeal region, whereas the basisphenoid floors the braincase in the posterior part of the nasopharyngeal region.

At the possible junction between the presphenoid and basisphenoid, a small splint of bone arises along the midline of the basisphenoid, here designated as a remnant of the parasphenoid. Immediately posterior to this splint but still at the anterior apex of the basisphenoid, there are one or two pits of unknown function. However, there is no indication of any foramen passing through the basisphenoid or close by.

The dorsal or endocranial surface of the presphenoid-basisphenoid is elevated anteriorly. The elevation is termed the tuberculum sellae (see also Hahn, 1981). In front of it is a shallow, anteriorly-curved, transverse groove, which lies immediately behind the optic foramina and which is designated herein as the chiasmatic sulcus. Additionally, a median shallow groove divides the tuberculum sellae symmetrically. Posteriorly, the presphenoid may or may not aid the basisphenoid in forming the anterior wall of the sella turcica (or pituitary fossa). This is uncertain because even in modern mammals the suture on the dorsal surface between the two bones is hardly observable. However, the sella turcica is present as a dominant feature on the endocranial surface of the basisphenoid.

The sella turcica is a deep fossa excavated into the basisphenoid, bounded laterally on each side by the medial bony wall of the cavum epiptericum (*i. e.*, the primary lateral wall of the braincase), and posteriorly by a rather low transverse ridge, here termed the dorsum sellae. The sella turcica is more or less rectangular in dorsal view, with a pair of foramina situated at its anterolateral corners. Each of the foramina pierces the medial wall of the cavum epiptericum near (but posterior to) the sphenorbital fissure, and is here designated as the metoptic foramen. An additional foramen is situated midway between these two metopic foramina, and is here called nutrient foramen. Posteriorly, a much larger foramen passes through the medial bony wall of the cavum epiptericum and connects the cranial cavity with the semilunar fossa (*i. e.*, the part of the cavum housing the trigeminal ganglion). The foramen is here termed the prootic foramen. The bony element between the metoptic foramen and the prootic foramen of the medial wall of the cavum is here called the taenia clino-orbitalis, or clinoid process (equivalent to the pila antotica in the chondrocranium).

In endocranial view, there are no clear sutural delineations between the presphenoid and orbitosphenoid, nor between the basisphenoid and the taenia clino-orbitalis. Furthermore, there is no indication of the existence of any foramen or groove on the dorsal surface of the basisphenoid, which might allow entrance of the internal carotid artery into the sella turcica.

Discussion

Without unambiguous sutural delineations, recognition of the limits of a particular bone is a matter of individual interpretation. Admitted or not, this situation is usually a rule rather than exception in the literature dealing with small skulls of early mammals. Unfortunately, not all workers have made the distinction explicitly between observation and interpretation. It is not surprising that many myths concerning phylogenetic inference based upon "good" cranial characters stemmed from taking subjective interpretations as objective facts. Therefore, it is desirable and often crucial to express clearly where one's observation stops and interpretation begins.

Problem of recognition of presphenoid, basisphenoid, and parasphenoid.—It is less a problem to recognize the basisphenoid than to recognize the presphenoid and parasphenoid. The basisphenoid, by bearing a sella turcica and forming the cranial floor in the middle region anterior to the basioccipital, is unmistakably identified in *Lambdopsalis*. The basisphenoid is a bone ossified from the chondrocranium and forming part of the cranial base throughout tetrapods. Although the sutural demarcation between the basisphenoid and the bone anterior to it in ventral view is not certain, it seems that there is an additional bone between the vomer and the basisphenoid in *Lambdopsalis*. This additional bone (here called the presphenoid) has generated many discussions and, in turn, many controversies (*e. g.*, Parrington and Westoll, 1940; Moore, 1981).

There have been several ways of dealing with this bone anterior to the basisphenoid in early mammals in general and in multituberculates in particular. For example, this cultriform bone has been treated as the cultriform process of the parasphenoid in *Triconodon mordax* (see Kermack, 1963) and in *Morganucodon* (see Kermack and others, 1981). However, as Kermack and others (1981, p. 63) stated: "In *Triconodon*, the basisphenoid, the parasphenoid, the pterygoid and the alisphenoid were completely fused to form a single bone, the sphenoid. Kermack's figures show that in BM. 47763 the basisphenoid itself is effectively complete. The parasphenoid is fused to the ventral surface of the basisphenoid: it is impossible to make out where one bone ends and the other begins." The same is claimed to be true of *Morganucodon* (*ibid.*). Initially, Kermack (1963) admitted that his reconstruction of a basisphenoid-parasphenoid complex in *Triconodon* was based upon Crompton's verbal suggestion of ". . . interpreting the bone in question not as if it were the sphenoid of a modern mammal but as if it resembled the sphenoid complex of an advanced mammal-like reptile" (Kermack, 1963, p. 94). With the same belief, Kermack and others (1981) described the sphenoid complex of *Morganucodon* in exactly the same way. Yet they expressed the view that: "There is a close resemblance between the basicranium of *Triconodon*, as reconstructed by Kermack (his fig. 14), and our reconstruction of the basicranium of *Morganucodon* (Fig. 98)" (Kermack and others, *ibid.*, p. 64). In this case, it would be surprising if there were not a "close resemblance"! Consequently, both Kermack (1963) and Kermack and others (1981) suggested that the retention of a parasphenoid in *Triconodon* and in *Morganucodon* is a reptilian character, which, together with a few other features, demonstrated their intermediate nature in the therapsid/mammal transition. Circular reasoning underlying their argument is thus evident.

On the other hand, while describing a basisphenoid-parasphenoid complex in the tritylodontid *Kayentatherium*, Sues (1986) nevertheless thinks that the complex includes ". . . a possible presphenoid component" (p. 236). Here the questions may arise: (1) whether the basisphenoid-parasphenoid complex in triconodonts as reconstructed by Kermack and his associates may also include a possible presphenoid component; and (2) if there has to be a clear distinction between a base of the sphenoid complex including parasphenoid but not presphenoid in advanced mammal-like reptiles and early mammals on one hand, and that including presphenoid but not parasphenoid in modern mammals on the other. After all, presphenoid is a cartilage bone whereas parasphenoid is a dermal bone, and thus one does not necessarily have to be replaced by the other.

Among multituberculates, Simpson (1937) first attempted to reconstruct the limits of the basisphenoid in *Ptilodus*, and labeled a cultriform bone immediately anterior to (but at a more ventral level than) the basisphenoid as the vomer. This practice has been essentially followed by Kielan-Jaworowska (1971) for some Mongolian taxa and Hahn (1981) for *Pseudobolodon*. Incidentally, this kind of arrangement is also seen in the adult skull of *Ornithorhynchus* (see Griffiths, 1978). However, the bone dorsal to the vomer and immediately anterior to (and at the same level as) the basisphenoid remains to be identified. In *Ectypodus*, Sloan (1979) avoided labeling the bone anterior to the basisphenoid. However, there seems to be no posterior extension of the vomer underlying the bone in question. More recently, Kielan-Jaworowska and others (1986) described a basisphenoid-presphenoid complex in *Chulsanbaatar*, although "The boundary between the basisphenoid and presphenoid cannot be identified with certainty" (p. 543). This condition is similar to that seen in *Lambdopsalis*. But *Lambdopsalis* has a proportionally longer exposure of the cultriform bone between the vomer and the basisphenoid. It is, therefore, tempting to follow Kielan-Jaworowska and others (1986) and interpret this cultriform bone in *Lambdopsalis* as presphenoid.

This interpretation is strengthened somewhat by the presence of a splint-like bone lying ventrally along the midline of the basisphenoid in *Lambdopsalis*; it could be accounted for as a remnant of the parasphenoid. A similar splint bone, occupying a similar position and identified as such, has been reported in *Galeopithecus* (= *Cynocephalus*) (see Parrington and Westoll, 1940) and *Didelphis* (see Fuchs, 1908; Parrington and Westoll, 1940; Jollie, 1962; Presley and Steel, 1978). Parrington and Westoll (1940, p. 339) argued strongly: "Considering the relations of this small bone, it is very remarkable that its claims to be the parasphenoid have been so often overlooked in favour of the extremely dubious claims of the vomer." Furthermore, it is clear that this splint bone in *Lambdopsalis* is separated from the vomer by a relatively long gap in which the presphenoid is exposed (as interpreted above).

Indeed, the presumed unaccountable disappearance of the parasphenoid in Mammalia as treated in standard textbooks of comparative anatomy of vertebrates (*e. g.*, Romer and Parsons, 1977) is an oversimplification. Various bones in the vicinity of the basisphenoid of mammalian skulls have been homologized, at least in part, with the parasphenoid. These include vomer (*e. g.*, De Beer, 1937), pterygoid (De Beer, 1929; and Goodrich, 1930), and even promontorium of the petrosal (Gow, 1985; this problem will be discussed under the section "Petrosal and Auditory Region"). Many controversies have surrounded these problems, a review of which is certainly beyond the scope of the present work (for good reviews, see Parrington and Westoll, 1940; Presley and Steel, 1978; and Moore, 1981). Nevertheless, it should be noted that, besides the examples cited above, a remnant of parasphenoid has been found in a number of mammals beneath the basicranium posterior to the vomer, which is supposed to arise ". . . as in reptiles, in the prochordal mesoderm of the stomatodaeum" (Green and Presley, 1978, p. 216). Therefore, it seems plausible to interpret the splint bone in *Lambdopsalis* as a remnant of the parasphenoid.

In summary:

1. There is no solid evidence to support or deny the assumption that the base of the sphenoid complex in early mammals is similar to that of therapsids rather than to that of modern mammals.

2. The presphenoid and parasphenoid coexist in a number of modern mammals. Among fossils, at least in *Lambdopsalis*, the base of the sphenoid complex may include components not only of a parasphenoid but also of a presphenoid.

3. The variations in the extent of different bones forming the base of the sphenoid complex seen within multituberculates alone suggest that there is no fixed pattern in a given group. Therefore, the reduction or even loss of certain bones (*e. g.*, parasphenoid) in the complex probably has doubtful significance in phylogenetic analysis.

Intracranial features of presphenoid—basisphenoid complex.—Although the ventral surface of the base of the sphenoid complex in multituberculates has been extensively described, relatively little is known about the features of its dorsal or intracranial surface. Hahn (1981) described the dorsal surface of the base of the sphenoid complex in a Guimarota multituberculate, *Pseudobolodon*. It represents the first and only adequate knowledge of anatomical details in that part of a multituberculate skull. The serial sections of Mongolian multituberculate skulls made by Kielan-Jaworowska should allow a detailed reconstruction of the intracranial features of the region as a whole, but this awaits publication.

The intracranial features of the presphenoid-basisphenoid complex in *Lambdopsalis*, as described earlier, are essentially the same as those of *Pseudobolodon* (see Hahn, 1981). The only major difference lies in the absence of a pair of internal carotid foramina in the posterior part of the basisphenoid in *Lambdopsalis*. Somewhat related to this difference is Hahn's (1981) description of lateral perforations of the dorsum sellae by a pair of foramina carotica in *Pseudobolodon*. However, there is a gap between the dorsum sellae and the medial bony wall of the cavum epiptericum on both sides in *Lambdopsalis*. The latter condition seems to be more characteristic of mammals in general. Even though in therian mammals the medial confines of the cavum epiptericum fail to ossify (but are marked by dura mater), the topographic relation between the dorsum sellae and the dura mater remains similar to that between the dorsum sellae and the osseous pila antotica as seen in *Lambdopsalis*. Other less significant differences between these two taxa may be more apparent than real. For example, the base of the sphenoid complex in *Pseudobolodon* seems to be proportionally narrower and more elongated than that of *Lambdopsalis*. This may well be caused by distortion of the skull of *Pseudobolodon* due to lateral compression. The lateral compression also may have caused the sella turcica of Hahn's specimen to become distorted into a triangular shape with a rather narrow anterior apex (rather than being rectangular as in *Lambdopsalis* and mammals in general). Consequently, two small foramina ("pores [ph]" *sensu* Hahn, 1981) at the bottom of the anterior apex of the sella turcica in *Pseudobolodon* are squeezed together, whereas they are at the anterolateral corners of the sella turcica in *Lambdopsalis*.

Despite the fact that the intracranial features of the base of the sphenoid complex are similar between *Pseudobolodon* and *Lambdopsalis*, I interpret some of these features rather differently from Hahn (1981):

1. Hahn (1981, p. 240) observed that, in *Pseudobolodon*, "2 small pores (ph) are present" at the bottom of the anterior apex of the triangular sella turcica. He went on: "As is seen on the ventral surface, these pores do not penetrate the basisphenoid" (*ibid.*, p. 240). He also excluded the possibility of their being remnants of the "fenestra hypophyseos," and was puzzled by their unknown function. However, a pair of small foramina are present in a similar position in *Lambdopsalis*; they pass anterolaterally through the medial walls of the cava epipterica, and herein are interpreted as the metoptic foramina. In the mammalian neurocranium, the metoptic foramen marks the boundary between the pila metoptica (anteriorly) and the pila antotica (posteriorly), and allows the oculomotor nerve, the ophthalmic artery, and the pituitary vein to pass laterally into the cavum epiptericum and then, via the sphenorbital fissure, into the orbit (De Beer, 1937). Therefore, in early mammals (such as multituberculates) in which the primary side wall of the braincase was ossified in adults, it would be expected that the metoptic foramen remained there for the passage of the nerve and blood vessels, just as seen in *Lambdopsalis* and *Pseudobolodon*. Even in adult modern mammals, in which the bony primary side wall of the braincase is either reduced (*e. g.*, in monotremes) or completely lost (*e. g.*, in therians), evidence for the perforations of the primary wall by the neural and vascular elements can be sought from the primary wall's close neighbor—the dura mater. The perforations of the dura mater by eye-muscle nerves and accompanying blood vessels suggest that the

topographic relations between the primary side wall and the neural and vascular structures remained unchanged throughout the therapsid/mammal transition. As De Beer (1937, p. 390) pointed out: "It was noticed by Sutton (1888) that in mammals the eye-muscle nerves perforate the dura mater in places which are not directly opposite their (eventually formed) foramina of exit from the skull. This remarkably acute observation can now be explained: the nerves perforate the dura mater as they did in the ancestral mammalian forms when the original side wall of the chodrocranium (pila antotica) was regularly chondrified." Conversely, this not only most satisfactorily explains the function of the two small foramina in *Pseudobolodon* but also explains why they "do not penetrate the basisphenoid" (Hahn, 1981, p. 240).

2. In summarizing his study of the sphenoid in *Pseudobolodon*, Hahn (1981) reached several conclusions, some of which may require further clarifications. For instance, Hahn (*ibid.*, p. 241-242) concluded that, in *Pseudobolodon*, "The cavum epiptericum is completely incorporated into the cranium as in adult monotremes . . . The border of the primary lateral wall is marked by the crest gr." (*i. e.*, Hahn's "Grat zwischen der Fossa hypophyseos und dem Cavum epiptericum" [= ridge between hypophyseal fossa and cavum epiptericum]).

According to De Beer (1937, p. 430), the cavum epiptericum in "lower vertebrates" ". . . is an extracranial space situated laterally to the side wall of the orbitotemporal region of the skull, dorsally to the basitrabecular process, and medially to the processus ascendens of the pterygoquadrate cartilage." In cynodonts, the cavum epiptericum remains extracranial (Presley, 1980). In monotremes, the cavum epiptericum is included within the bony skull, but its medial wall, although reduced, remains as a bony ridge, the taenia clino-orbitalis (= ossified base of the pila antotica) (Goodrich, 1930; Jollie, 1962). In marsupials and eutherians, the cavum epiptericum is completely incorporated into the bony skull ". . . since the pila antotica is no longer present, and the boundary between the cavum and the cranial cavity is indicated only by the dura mater" (De Beer, 1937, p. 430). In the Late Cretaceous Mongolian multituberculates (Kielan-Jaworowska and others, 1986) and in *Lambdopsalis*, the cavum epiptericum is also included in the bony skull, but its medial bony wall remains little reduced, and is much more extensive than that of monotremes.

As to *Pseudobolodon*, Hahn (1981) reconstructed the medial wall of the cavum epiptericum as marked by the "crest gr.", and implied that this condition is exactly the same as that of monotremes. If this is indeed the case, then we have to assume that *Pseudobolodon* is more derived in this aspect than other more advanced multituberculates. To overcome the contradiction of this character against many other characters, we have to assume that the reduction of the taenia clino-orbitalis in *Pseudobolodon* is an autapomorphy among multituberculates, and perhaps independently developed. It seems very unlikely. However, more careful study of Hahn's (1981) paper reveals that the specimen of *Pseudobolodon* available to Hahn was naturally broken to expose the intracranial features, and that the dorsal limit of the "crest gr." was poorly defined. In other words, unless it can be shown that the dorsal surface of the "crest gr." as preserved in the specimen is a natural ending of the dorsal extension of the bone rather than a break, the possibility remains that there is a substantially complete bony medial wall of the cavum epiptericum in *Pseudobolodon*, just as that seen in *Lambdopsalis* and other Mongolian multituberculates. If this is true, there is no need to assume an independent reduction of the medial bony wall of the cavum in *Pseudobolodon*.

In turn, the fact that the bony, primary side wall of the braincase is still fairly complete in multituberculates (even though the secondary side wall also has been completed) implies that their braincase structure seems to be more primitive than that of monotremes. This is also contrary to Hahn's (1981) conclusion that "the cavum cranii of the multituberculates is . . . not less evolved than that of the monotremes" (p. 242).

In summary, the cavum epiptericum in multituberculates is a space between the primary and secondary walls of the braincase, and therefore provides a morphological intermediate between that of reptiles and that of modern mammals.

Problem of cavum epiptericum in early mammals.—Since the incorporation of the cavum epiptericum into the braincase has been assumed to be a mammalian feature, the structure of the cavum in advanced cynodonts and early mammals has been extensively discussed (*e. g.*, Goodrich, 1930; De Beer, 1937; Kermack, 1963; Hopson, 1964; Presley and Steel, 1976; Kuhn and Zeller, 1987; and Maier, 1987). However, differences in understanding, or more often in terminology, of the surrounding structures of the cavum epiptericum have troubled communications among researchers.

Kermack (1963) stated: ". . . the Triconodonta had a braincase of an essentially reptilian pattern. There was a persistent cavum epiptericum lying outside the ossified lateral wall (formed by the petrosal) of the braincase" (p. 83). Kermack (*ibid.*) believed that this was also true of *Morganucodon*. This interpretation would have raised no difficulty, had Kermack (*ibid.*) not gone on to suggest that the anterior lamina of the petrosal is "an ossification within the wall of the neurocranium itself" (p. 97), and that "the cavum epiptericum would finally vanish as the anterior lamina of the petrosal came into contact with the alisphenoid" (p. 97). Apparently these statements are self-contradictory, because the anterior lamina of the petrosal cannot be both an ossification of the neurocranium and a part of the secondary lateral wall (after making contact with the alisphenoid). Similarly, the anterior lamina can hardly be both lateral and medial to the cavum epiptericum. This led Hopson (1964) to suppose that Kermack interpreted the cavum epiptericum "as having been subdivided by the formation of the anterior lamina, with part of it enclosed within the neurocranium and part of it left outside of the braincase proper but still medial to the alisphenoid" (Hopson, 1964, p. 22). This is most unlikely to be true, if one considers the topographic relations among the endocranial cavum epiptericum, the neurocranium, and the braincase in both monotremes and modern therians. Consequently, Hopson (*ibid.*) offered a more reasonable interpretation, suggesting that "the cavum epiptericum lay entirely *medial* to both the anterior lamina of the periotic (posteriorly) and the alisphenoid (anteriorly)" (p. 24, italics original). He also noted that the ossified wall medial to the cavum epiptericum is the retention of the ossified pila antotica. Hopson (1964) further pointed out that origin of the anterior lamina of the periotic of monotremes "as an intramembranous ossification strongly suggests a similar mode of origin for the anterior lamina of both tritylodonts and Mesozoic mammals" (p. 23).

However, Kermack and others (1981) insisted, in their conjectural transverse sections of the skull of *Morganucodon*, that the cerebral dura mater (adjacent to the pila antotica in the chondrocranium) "divided the cavum epiptericum into two parts: a ventral region that lay outside the cranial cavity (c. ep.) and a dorsal region that had become incorporated into the braincase" (p. 126). They also claimed: "This is typical of the Mammalia" (*ibid.*, p. 126). In other words, they see the cavum epiptericum of *Morganucodon* as essentially an unfloored extracranial space plus an endocranial part medial to the dura mater in life. To the best of my knowledge, this condition, if true, would be unique among mammals. Kermack and others' claim of its being typical of mammals has no factual basis.

More recently, Crompton and Sun (1985) interpreted that "in *Morganucodon* and *Sinoconodon* the anterior lamina forms the medial wall and a floor to the cavum develops between the lateral flange and the dorsolateral edge of the greatly enlarged promontorium (cochlear housing)" (p. 109). This interpretation differs from that of Kermack and others (1981) in adding a floor to the cavum epiptericum, but agrees with the latter on the anterior lamina being the medial wall of the cavum. Again, Crompton and Sun's interpretation is contrary to Hopson's (1964) interpretation that the cavum epiptericum is *medial* to both the anterior lamina and the alisphenoid. Incidentally, the interpretation offered by Crompton and Sun (1985), like that of Kermack and others (1981), also demands an additional anterior lamina of the petrosal to lie lateral to the cavum epiptericum. As such, Crompton and Sun (1985) stated: "The pro-otic has grown forward as a lateral lamina to surround the two branches of the fifth nerve and form a lateral wall to the cavum epiptericum" (p. 109).

This confusing picture can be understood better if Hopson's (1964) interpretation is accepted. In addition, although the medial wall of the cavum epiptericum is believed to be "marked posteriorly by the pila antotica arising from the prootic" (Presley, 1980, p. 159), De Beer (1937) stressed that the reason why he preferred the term "pila antotica" to the "pila prootica" is that the latter is "liable to convey the false impression that the structure has anything to do with the prootic bone" (p. 388). Clearly, the kind of "false impression" has been hardly avoided even by using the term "pila antotica." I suspect that many of the above-discussed confusions may be attributed to the misnomer of the ossified pila antotica as the anterior lamina of the petrosal. This misnomer was also inherited in Moore's (1981) otherwise excellent review of the mammalian skull, in which the ossified pila antotica was equated with the anteroventral process of the petrosal.

Because neither the medial wall (*i. e.*, ossified pila antotica) nor the lateral wall (*i. e.*, "anterior lamina of petrosal" in conventional sense) of the cavum epiptericum is a true outgrowth from the petrosal (see also Hopson, 1964; Griffiths, 1978; Presley, 1981; and later discussion in the present work), the anterior lamina of the petrosal is a misleading term and should be discarded. However, Gow (1985, p. 136) reported "the presence of neurocranial anterior lamina" in *Tritylodon*. Gow (*ibid.*, p. 146) interpreted that "the epipterygoid is sandwiched between processes of prootic," and that the medial process "must be considered to be neurocranial bone" and the lateral process "can be considered a membrane bone component." It is equally parsimonious to argue that the medial process might be equivalent to the ossified pila antotica and the lateral process to the lamina spheno-obturans.

I also think that the expressions of the cavum epiptericum as an extracranial space in cynodonts versus its endocranial nature in modern mammals are equally misleading. When we say: "The cavum epiptericum is extracranial in cynodonts" (Presley, 1980, p. 159), we are saying that the cavum is lateral to the primary side wall of the braincase. Similarly, when we say: "All Recent mammals have an endocranial cavum epiptericum" (Presley, *ibid.*, p. 160), we are actually saying that the cavum is medial to the secondary side wall of the braincase whereas the primary side wall fails to ossify (but is marked by the membranous dura mater). The topographic relation between the surrounding structures of the cavum epiptericum remains unchanged. In all cases, the cavum is extracerebral.

With all these in mind, I attempt to interpret the surrounding structures of the cavum epiptericum in cynodonts and mammals as follows:

1. In cynodonts, the cavum epiptericum lies outside the ossified primary lateral wall of the neurocranium. The epipterygoid is a bony bar lateral to the cavum, but no complete secondary lateral wall has been formed (nor has the floor to the cavum).

2. In multituberculates, and probably triconodonts, the secondary side wall of the braincase has been completed, and the pila antotica remains completely ossified. Thus, the cavum epiptericum is between these two bony walls. Consequently, certain cranial nerves and accompanying blood vessels pass through both of the bony walls. For example, as described and discussed earlier, in *Lambdopsalis* the oculomotor nerve, the ophthalmic artery, and the pituitary vein pass through the metoptic foramen on the primary side wall into the cavum epiptericum, and then pass into the orbit through the sphenorbital fissure on the secondary lateral wall. Similarly, the trigeminal nerve and the abducens nerve pass through the

prootic foramen on the primary side wall, and then into the cavum. Whereas the trigeminal nerve may pass through several foramina on the secondary side wall, the profundus branch of the trigeminal nerve as well as the abducens nerve pass into the orbit through the sphenorbital fissure.

3. In monotremes, the secondary lateral wall of the braincase is complete, but the primary bony wall is reduced to a low ridge (the ossified base of the pila antotica), which still defines the medial limit of the cavum epiptericum (De Beer and Fell, 1936).

4. In marsupials and eutherians, an ossified pila antotica has completely disappeared, and the primary lateral wall of the neurocranium is marked by cerebral dura mater. However, the space between the dura mater and the secondary side wall of the braincase, *i. e.*, the cavum epiptericum, remains, and the dura mater bears exactly the same relation with certain cranial neural and vascular structures as the ossified pila antotica in more primitive forms (see Goodrich, 1930, Fig. 280).

Therefore, the general trend in evolutionary changes of this part of the braincase from cynodonts leading toward modern therians seems to show the completion of the secondary lateral wall and reduction (and finally the disappearance) of the ossified primary lateral wall of the braincase. If this interpretation is to be accepted, Kermack's (1963, p. 83) statement that "the Triconodonta had a braincase of an essentially reptilian pattern" would be only partly true. That is, the triconodonts' braincase is "reptilian" in terms of the probable retention of a complete ossified medial wall of the cavum epiptericum, and is mammalian in terms of the completion of the secondary side wall. Furthermore, as documented clearly in *Lambdopsalis*, multituberculates retained a completely ossified medial wall of the cavum epiptericum, and thus at least in this aspect seem to be more primitive than monotremes.

My discussion to this point has depended upon the widely-accepted assumption that the cavum epiptericum is an essentially homologous structure throughout the taxa concerned. Recently, however, Kuhn and Zeller (1987) suggested that such presumed homology may not exist. They (*ibid.*, p. 62) claimed that "the extent and contents of the cavum epiptericum differ markedly among three groups of mammals" (*i. e.*, *Tachyglossus, Ornithorhynchus*, and modern therians). They also challenged the homologies of the bony elements of the lateral wall of the cavum epiptericum between monotremes and therian mammals. Therefore, they (*ibid.*, p. 64) concluded that the cavum epiptericum "was formed independently in monotremes and therians during phylogeny." This problem will be discussed further in connection with the problem of the origin of mammalian alisphenoid.

Problem of internal carotid foramen.—Simpson (1937) first described a pair of carotid foramina within the posterior part of the basisphenoid in *Ptilodus*. He also described two pairs of tiny nutritive foramina anterior to the carotid foramina. Kielan-Jaworowska (1971) identified two minute foramina (situated at the positions comparable to the anterior pair of the nutritive foramina in *Ptilodus*) as carotid foramina in *Kamptobaatar*. However, Kielan-Jaworowska and others (1986, p. 579) pointed out that the real carotid foramen was "previously misidentified as opening in floor of cavum epiptericum in *Kamptobaatar* (Kielan-Jaworowska, 1971)." (As an aside, Hahn, 1981, interpreted the opening in the floor of the cavum epiptericum in *Kamptobaatar* as foramen pseudovale inferium). This emended reconstruction by Kielan-Jaworowska and others (1986) places the carotid foramen anterior to the promontorium and within the suture between the basisphenoid and petrosal. They also regarded the course of the carotid canal in the Late Cretaceous Mongolian multituberculates as being similar to that in therian mammals. They further noted that the carotid foramen probably pierces the basisphenoid in Jurassic multituberculates. This refers to Hahn's (1981) identification of carotid foramen within the basisphenoid in *Pseudobolodon*. The carotid foramen figured in Sloan's (1979) reconstruction of the skull of *Ectypodus* is shown to be in essentially the same position as that seen in the Mongolian multituberculates. As described above, except one or two tiny pit(s) at the anterior end of the basisphenoid (reminiscent of the anterior pair of the nutritive foramina in *Ptilodus*), there is no indication of the presence of the carotid foramen on either the ventral or dorsal surface of the basicranium in *Lambdopsalis*.

It is generally accepted that in mammal-like reptiles, early nontherians (*e. g.*, triconodonts, multituberculates and monotremes), and most marsupials, "The medial branch of the internal carotid artery passed through the entocarotid canal in the basisphenoid" (Marshall, 1979, p. 376; see also Patterson, 1965; Archibald, 1977). Wible (1986), however, argued forcefully that there is no correlation between a medial route for the internal carotid artery and the perforation of basisphenoid by the carotid foramina. The above review of various positions of the carotid foramen among different members within the order Multituberculata further suggests that no uniform position of the carotid foramen for the so-called "nontherians" exists.

Wahlert (1974, p. 373) pointed out that in *Marmota* the carotid canal "transmits a vein, the inferior petrosal sinus, which joins the internal jugular vein," but not the internal carotid artery as would be expected. Moreover, the fact that Matthew's (1909) dual-carotid hypothesis has been discredited recently (Presley, 1979; Wible, 1983, 1984, 1986; and MacPhee and Cartmill, 1986) also demonstrates that speculation about soft anatomy based upon purely osteological evidence should be done cautiously.

Furthermore, the high degree of diversity in position of the carotid foramen among multituberculates also reveals that the cephalic arterial pattern may vary considerably, a situation probably parallel to that in modern rodents (see also Bugge, 1974, 1985). The absence of the carotid foramen in *Lambdopsalis* may indicate the probable absence of the internal carotid artery, analogous to lack of the artery in many modern rodents (*e. g.*, protrogomorphs, sciuromorphs, and the Old World hystricomorphs etc.; Bugge, 1974). In these modern rodents, the role of the internal carotid artery is replaced by other arteries, especially the vertebral artery (Bugge, 1974). As a pure speculation, *Lambdopsalis* may well have independently developed a similar alternative for the arterial supply to the anterior part of the brain.

To sum up, the variations in presence versus absence of, and, if present, the position of, the carotid foramen shown in multituberculates strengthen Wible's (1986) argument that there is no *a priori* reason to assume the plesiomorphous state of the internal carotid artery necessarily associated with the primitive mammals. In addition, the fact that the traits related to the carotid foramen and internal carotid artery vary even within a single order cautions against their unbridled use in higher-level phylogenetic analysis of Mammalia.

Orbitosphenoid (Figs. 4, 5, 11, 15-17)

Description

The orbitosphenoid is recognized in the following specimens: IVPP V7151.2, V7151.51, V7151.52, V7151.77, V7151.81, V7151.85, V7151.90, V7151.92, and V7151.93.

The orbitosphenoid is a small, more or less vertical bone surrounding the optic foramen. It is in sutural contact with the orbital process of the frontal anteriorly, with the ascending process of the maxilla ventrally, with the alisphenoid posterodorsally, and with the ossified pila antotica posteriorly. The sutural delineation between the orbitosphenoid and the orbital process of the frontal is clearly visible in IVPP V7151.51 and V7151.85, located about half-way between the ethmoid foramen and sphenorbital fissure. The optic foramen is large, and situated in the posterior half of the orbitosphenoid bone, at a point directly below the postorbital foramen.

Immediately posteroventral to the "orbital ridge" (see also the description in section "Frontal"), the orbitosphenoid makes a short contact with the alisphenoid posterodorsally. However, much of the orbitosphenoid ventral to that contact extends medial to the alisphenoid, and makes major sutural contact with the ossified pila antotica posteriorly. The opening immediately ventrolateral to the articulation between the orbitosphenoid and the alisphenoid is the sphenorbital fissure. The sphenorbital fissure is posterolateral to the optic foramen, and can be seen only in an anterior view of the skull.

In lateral view, the suture between the orbitosphenoid and alisphenoid starts at the anterodorsal rim of the optic foramen. The alisphenoid screens the posterior part of the orbitosphenoid. In endocranial view, the suture between the orbitosphenoid and the ossified pila antotica runs posterodorsally in the area half-way between the optic and metoptic foramina. The orbitosphenoid overlaps the pila antotica medially.

Discussion

In mammals in general, the orbitosphenoid is an ossification of the chondrocranium, representing the cartilaginous pila metoptica and parts of the pila preoptica and taenia parietalis (Goodrich, 1930; De Beer, 1937; Oelrich, 1956; and McDowell, 1958). Although Gregory (1910) suggested that the orbitosphenoid probably was not perforated originally by the optic foramen, it appears that the orbitosphenoid is always formed in the vicinity of the exit for the optic nerve among mammals (McDowell, 1958).

Reconstruction of orbitosphenoid in multituberculates.—Until now, few known skulls of multituberculates show unambiguous sutural demarcations between the orbitosphenoid and its neighboring bones. As a result, reconstructions of the orbitosphenoid in various multituberculates are essentially authors' interpretations. For example, in *Ptilodus*, Simpson (1937, p. 745) stated: "Above the maxillary process containing the roots of M1 there appears to be a large, discrete, roughly quadrate, vertical, anteroposteriorly elongated element, perhaps the orbitosphenoid." Kielan-Jaworowska (1971) interpreted that in *Kamptobaatar* "The orbitosphenoid has a very large exposure in the orbit, occupying most of the interorbital wall" (p. 12). Consequently, her reconstruction made the orbitosphenoid in *Kamptobaatar* unusually large. Although the orbitosphenoid is also poorly preserved in other Mongolian taxa, Kielan-Jaworowska and others (1986) believed that the shape and size of the orbitosphenoid in them are probably as interpreted for *Kamptobaatar*. As I argued earlier, size of the orbit of the Mongolian multituberculates may well have been exaggerated significantly. This may explain why Kielan-Jaworowska and her associates conceived such a large orbitosphenoid as necessary. Sloan (1979) reconstructed proportionally a much smaller orbitosphenoid in *Ectypodus*. However, the dashed line in Figure 1 of Sloan (1979) representing the sutural delineation between the orbitosphenoid and its neighbors indicates that the confines of the bone are purely conjectural.

The fairly complete sutural demarcation between the orbitosphenoid and its neighbors in *Lambdopsalis* firmly suggests that, at least in this genus of multituberculates, the orbitosphenoid is not a large bone within the orbital mosaic. Although a large orbitosphenoid bone has been assumed to be a primitive feature in mammals (*e. g.*, Gregory, 1910; Muller, 1935; and Novacek, 1980), there is no *a priori* reason to presume that this character state in multituberculates should necessarily be primitive. In fact, even in some advanced cynodonts, the orbitosphenoid is not necessarily a dominant element in the orbit (Sues, 1986).

The orbitosphenoid in *Morganucodon* reconstructed by Kermack and others (1981) is also questionable. Although they did not define the posterior limit of the bone, the reconstructed part of the orbitosphenoid is already an enormous one. It is difficult to conceive that the anterior edge of the orbitosphenoid in *Morganucodon* extends that far forward, lying dorsal to the sphenopalatine foramen. This would be an extremely unusual case in mammals. However, both my personal observations and their figures suggest that no sutural contacts are visible in the specimens. Therefore, their interpretation of that part of the orbit as orbitosphenoid in *Morganucodon*

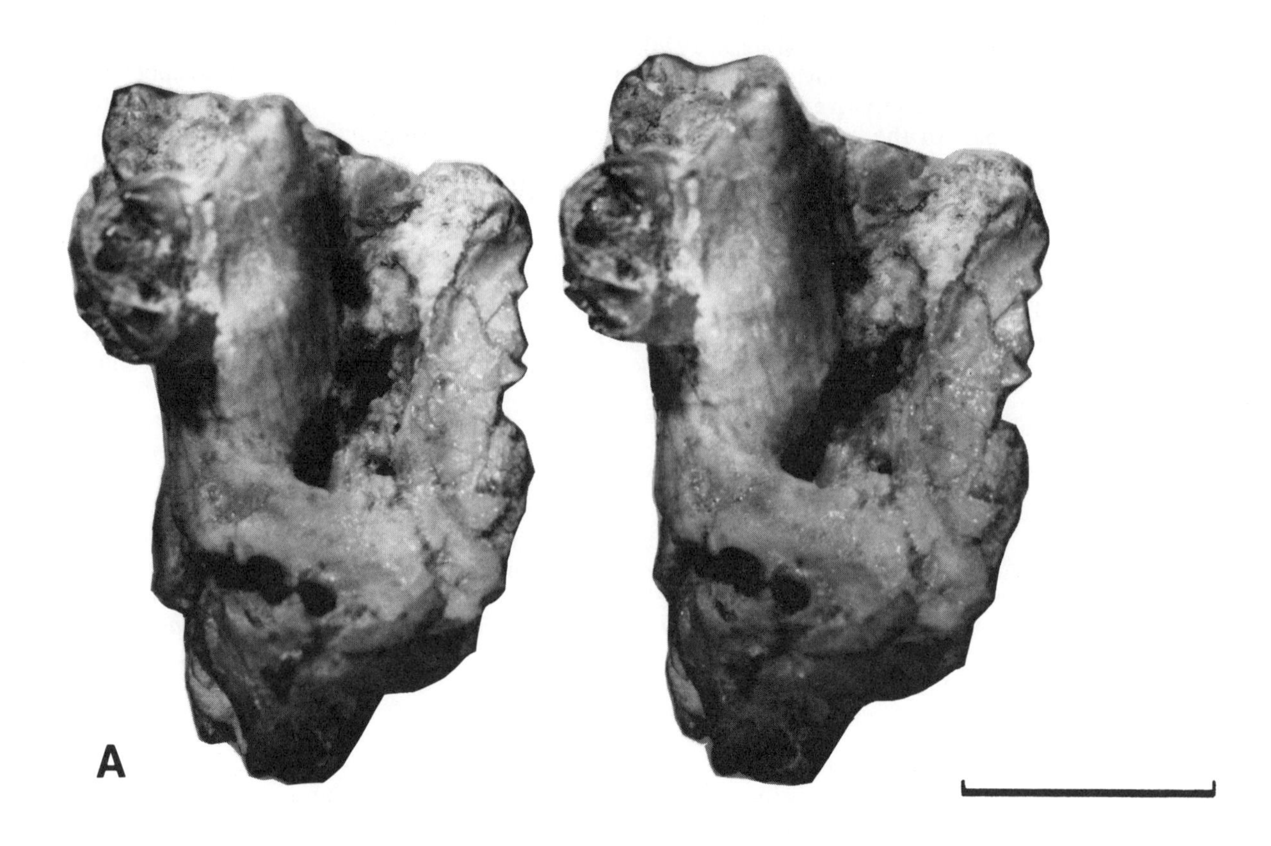

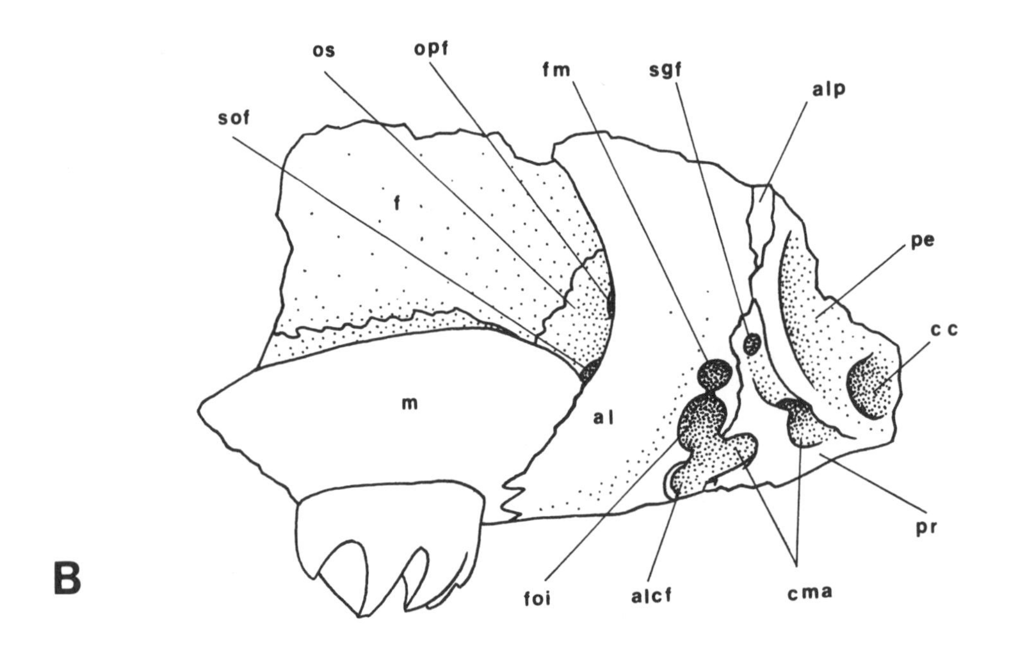

Figure 16. ***Lambdopsalis bulla*****, V7151.92, lateral view of fragment of skull.** ***A*****, stereophotograph, anterior to top; and** ***B*****, explanatory drawing, anterior to left.**

is tenuous. Nevertheless, it should be admitted that an objective identification of the confines of the individual bony element within the orbitotemporal region in recent mammalian skulls often proves to be difficult.

A striking feature in the orbital region of advanced cynodonts and early mammals that has been noticed by many workers (*e. g.*, Sues, 1986, for *Kayentatherium*; Kermack and others, 1981, for *Morganucodon*; Sloan, 1979, for *Ectypodus*; and Kielan-Jaworowska and others, 1986, for a variety of Late Cretaceous Mongolian multituberculates) is that the orbitosphenoid lies considerably medial to the alisphenoid (or the epipterygoid in cynodonts). This is also true of *Lambdopsalis*. Although Kielan-Jaworowska and others (1986) provided an implicit explanation for this observation only for multituberculates, it is certainly applicable to the similar feature in the other groups just discussed as well. They pointed out that the orbitosphenoid in multituberculates is "Continuous posteriorly below anterior lamina of petrosal with medial wall of cavum epiptericum as taenia clino-orbitalis" (p. 591). In other words, the cavum epiptericum in these groups has been (or is in the process of being) incorporated into the braincase, but the medial wall of the cavum is still retained and formed by the posterior part of the orbitosphenoid and the taenia clino-orbitalis (or ossified pila antotica). The ventral part of the alisphenoid forms the anterior part of the cavum's lateral wall. The wider the cavum, the more medial in position the orbitosphenoid. Only after the disappearance of the medial bony wall of the cavum epiptericum does the orbitosphenoid begin to have extensive posterior contact with the alisphenoid. It is interesting to note that, even without proper knowledge in embryology, this gross topographic relation between the orbitosphenoid and alisphenoid would help us understand that the orbitosphenoid is "an ossification of chondrocranium proper (in contrast to the alisphenoid, an ossification of the palatopterygoid cartilage)" (McDowell, 1958, p. 126).

It is also of interest to note that Simpson (1937) did not stress this striking feature in *Ptilodus*, nor was he aware of the retention of the cavum epiptericum in it.

Optic foramen in Lambdopsalis.—Simpson (1937) was unable to recognize a separate optic foramen in *Ptilodus*, and suggested that the optic foramen may be little (or not) differentiated from the sphenorbital fissure. But he pointed out that the foramen anterior to the sphenorbital fissure in *Ptilodus* cannot be the optic foramen because "it is far anterior to the postorbital constriction (which is always the approximate point of origin of the optic nerve)" (Simpson, 1937, p. 745-746). This observation also can be used to test Hahn's (1981) identification of the optic foramen in *Pseudobolodon*. It turns out that what Hahn called "optic foramen" in *Pseudobolodon* seems also to be far anterior to the postorbital constriction. In addition, Hopson (*personal communication*) informed me that the actual foramen in the specimen is much smaller than that in Hahn's illustration (Hahn, 1981, Fig. 4). Kielan-Jaworowska and others (1986) did not question Hahn's identification of the optic foramen in *Pseudobolodon*, but went further to suggest the possible presence of an optic foramen in *Nemegtbaatar*. However, in *Lambdopsalis*, with the foramen situated directly below the postorbital foramen combined with the chiasmatic sulcus immediately posterior to the foramen, this foramen is doubtless an optic formen. This should be considered as the first undoubted documentation of a separate optic foramen in a multituberculate.

However, I should stress that this does not imply that the other multituberculates might not have a separate optic foramen. In fact, the presence of a separate optic foramen should be defined by the presence of the ossified pila metoptica, not by the criterion of whether or not the optic foramen can be seen in lateral view. McDowell (1958, p. 126) pointed out: "Sometimes the foramen is present and defined by the pila metoptica, but lies deep within the foramen lacerum anterius (*e. g.*, most Soricidae), and is then said to be absent by some authors." In other words, in such a case, the optic foramen is concealed laterally by the alisphenoid, but in medial view the pila metoptica is present and the optic foramen is not confluent with the sphenorbital fissure. Therefore, one should be cautious about the conclusion of presence versus absence of the optic foramen when only the lateral view of the orbitotemporal region is available for observation.

Among living mammals, it has been assumed that a separate optic foramen occurs only in eutherians. The optic foramina in monotremes and marsupials are confluent with the sphenorbital fissure. Consequently, a separate optic foramen has been implicitly treated as an eutherian apomorphy (*e. g.*, W. K. Parker, 1885; Gregory, 1910; and Marshall, 1979). For example, W. K. Parker (1885, p. 271) stated: "the absence of a special optic foramen, is of similar import; there never is such a foramen until we are among the Placental Mammalia." Gregory (1910, p. 429) also mentioned: the optic nerve "originally issued through the sphenorbital fissure (foramen lacerum anterius) (*cf.* Monotremata, Marsupialia, *Sorex*). It later pierced the orbitosphenoid (most Placentals)." In his cladistic analysis of metatherian and eutherian characters, Marshall (1979) seemed to conduct an outgroup comparison to strengthen this point by stating: "In marsupials and *reptiles* the optic foramen and foramen lacerum anterium (sphenoidal fissure) are confluent and the nerve does not pierce the cartilaginous orbitosphenoid as it does in placentals" (p. 404, italics mine). In fact, a separate optic foramen defined by the pila metoptica is present in many living reptiles, *e. g.*, lizards (Oelrich, 1956; Starck, 1979; and Bellairs and Kamal, 1981), and *Sphenodon* (see Romer, 1956). It is also present in many fossil reptiles (see Romer, 1956, for details). More recently, Sues (1986, p. 237) described in *Kayentatherium* "N. opticus (II) emerged through a short canal between the anterior margin of the epipterygoid and the orbitosphenoid above the fissura orbitalis." Clearly, contrary to the belief of traditional wisdom, a separate optic foramen may represent a primitive character rather than a derived one in mammals.

The polarity of presence versus absence of the optic foramen would be clearer if one takes into account developmental process of the skull. As already discussed above, the optic foramen is defined by the pila metoptica. According to Moore (1981) and Kuhn and Zeller (1987), the reptilian ancestor of mammals has both the pila metoptica and pila antotica, and "Independently, monotremes have lost the Pila metoptica, whereas eutherians (placental mammals) have lost the Pila antotica. Marsupials are the most advanced in that they have lost both the Pila metoptica and the Pila antotica" (Kuhn and Zeller, 1987, p. 51). It already has been noted that *Lambdopsalis* retained both the ossified pila metoptica and pila antotica. Therefore, *Lambdopsalis* retained an essentially primitive condition in this aspect, and is thus more primitive than monotremes. Apparently, it is more parsimonious to assume that the presence of a pila metoptica and, in turn, a separate optic foramen in *Lambdopsalis* and eutherians is a symplesiomorphy rather than a synapomorphy or a homoplasy.

Contrary to the views held by Moore (1981), Kuhn and Zeller (1987), and others, Maier (1987) believes that the pila metoptica of reptiles and mammals are not likely to be homologous structures. His belief was based upon: "Neither mammal-like reptiles, nor the most primitive mammalian groups (Morganucodontidae; Multituberculata; Monotremata; Marsupialia) possess a separate optic foramen" (Maier, 1987, p. 87). From the above discussion, it is clear that Maier was partly misinformed. As mentioned earlier, only monotremes and marsupials definitely lack a separate optic foramen, and incidentally we do not know whether or not a separate optic foramen exists in Morganucodontidae (see Kermack and others, 1981). As such, Maier's challenge would raise no difficulty at all to Kuhn and Zeller's interpretation, especially in light of the discovery of a separate optic foramen in *Lambdopsalis*. Moreover, their interpretation is more parsimonious than Maier's, because the latter must involve an independent development of the pila metoptica in eutherians.

As an aside, although Jollie (1962, Figure 3-8, C) illustrated a separate optic foramen in *Ornithorhynchus*, it may well be what De Beer and Fell (1936) called the "pseudoptic foramen." "It represents the conjoint optic and metoptic foramina of other forms, confluent owing to the absence of a pila metoptica" (De Beer and Fell, 1936, p. 10).

In summary, the presence of a separate optic foramen in *Lambdopsalis* appears to strengthen Kuhn and Zeller's argument that reduction of the primary wall of the braincase in the earliest mammalian groups such as multituberculates may not have proceeded very far, and that the losses of various individual pillars of the primary wall of the braincase occurred independently later in monotremes, marsupials, and eutherians. This seems to imply that the phylogenetic separations among multituberculates, monotremes, and therians may have occurred very early in mammalian history.

Sphenorbital fissure.—The sphenorbital fissure in *Lambdopsalis* is essentially the same as that in the Late Cretaceous Mongolian multituberculates described by Kielan-Jaworowska and others (1986) as well as in the other described multituberculate skulls (Simpson, 1937; and Hahn, 1981).

One relevant difference is that, in *Lambdopsalis*, the optic nerve emerged through a separate optic foramen anteromedial to the sphenorbital fissure; in other multituberculates, the optic nerve passed into the orbit via the sphenorbital fissure. In the latter case, the optic foramen may be absent, or it may be present but lying deep within the sphenorbital fissure, concealed by the alisphenoid. Therefore, the sphenorbital fissure in *Lambdopsalis* allowed passage of most of the contents of the cavum epiptericum forward into the orbit (*e. g.*, the oculomotor, trochlear, profundus, and maxillary branches of the trigeminal, and abducens nerve). Again similar to the Mongolian taxa, the sphenorbital fissure in *Lambdopsalis* also, ". . . forms the anterior opening of the cavum epipterícum and can be seen, by looking backwards into the posterior prolongation of the temporal fossa" (Kielan-Jaworowska and others, 1986, p. 544). This condition is different from that seen in modern therians, in which the sphenorbital fissure is a foramen between the orbitosphenoid and alisphenoid. Apparently, this is due to the retention of a medially bony walled cavum epipterícum in multituberculates, among other early mammalian groups, which prevents the orbitosphenoid from articulating with the alisphenoid in most parts.

Another significant difference is that in the Late Cretaceous Mongolian taxa, the lateral wall of the sphenorbital fissure is described by Kielan-Jaworowska and others (1986) as "the anterior, crescentic margin of the anterior lamina" (p. 553) of the petrosal, whereas the lateral wall of the sphenorbital fissure in *Lambdopsalis* is undoubtedly the alisphenoid. This supposed difference, however, may turn out to be more apparent than real if the sutural delineations between the various bony elements of the orbitotemporal region in Mongolian forms would be visible. This problem will be discussed further in the following section.

Alisphenoid (Figs. 3-5, 11, 13-19)

Description

The alisphenoid is preserved in the following specimens: IVPP V5429, V5429.15, V7151.2, V7151.5, V7151.51-53, V7151.61-62, V7151.74, V7151.77, V7151.79, V7151.81, and V7151.90-94.

The alisphenoid can be divided arbitrarily into two parts. The line of division is chosen at the level of the dorsal rim of the optic foramen, where the alisphenoid makes short contact with the orbitosphenoid (see also the description in section "Orbitosphenoid"). Below this arbitrary line, the dorsal half of the alisphenoid forms the lateral wall of the cavum epipterícum. The anterior margin of this part of the alisphenoid is a free edge, laterally bordering the sphenorbital fissure and making no contact with the orbitosphenoid. The ventral half of the part of the alisphenoid below the arbitrary line has a long anterior process and a slightly shorter posterior process. The anterior process is in sutural contact with the maxilla anteriorly, with the nasal plate of the palatine antero-

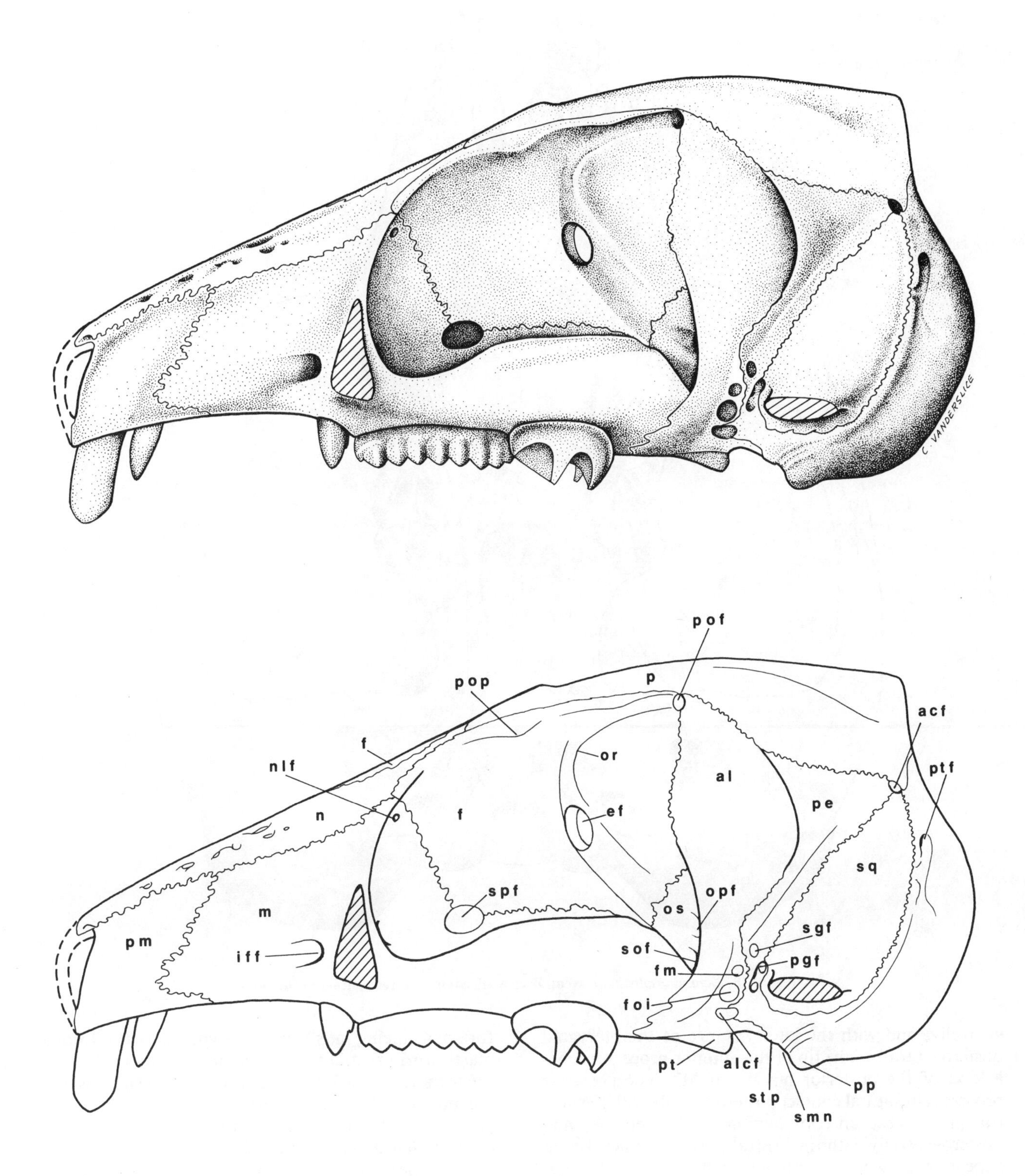

Figure 17. ***Lambdopsalis bulla*****, lateral view of composite reconstruction of skull.**

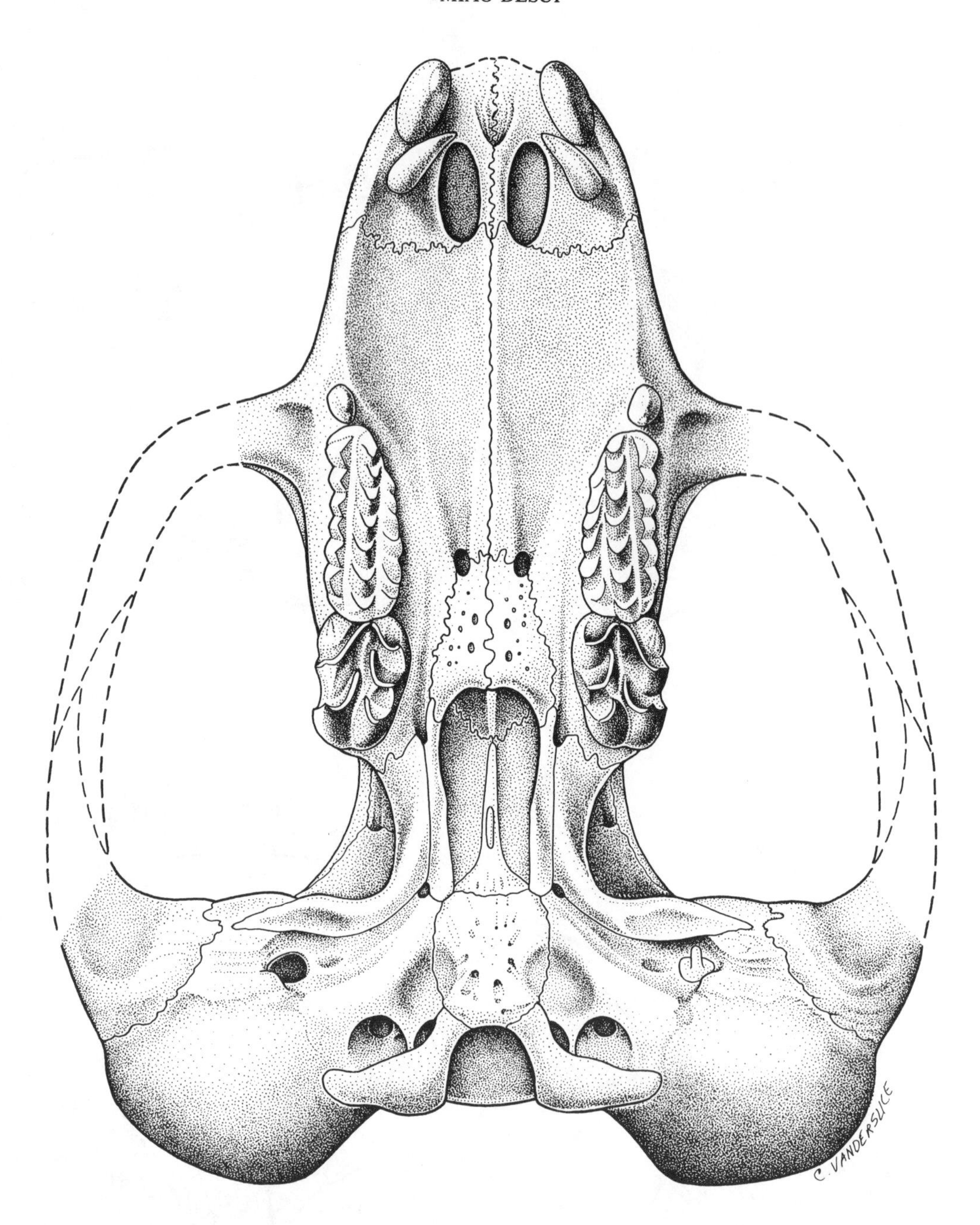

Figure 18. ***Lambdopsalis bulla*, ventral view of composite reconstruction of skull.**

ventrally, and with the lateral flange of the pterygoid posteroventrally. The tip of the anterior process reaches a level of the posterior margin of M2. The posterior process is in sutural contact posterodorsally with the ventral part of the anterior lamina of the petrosal, and posteroventrally with the ventral part of the petrosal at a location adjacent to the anteromedial rim of the fenestra vestibuli. The tapering end of the posterior process of the alisphenoid is defined clearly by the suture in specimen IVPP V7151.94.

In the posteroventral corner of the part of the alisphenoid below the arbitrary line, there are two foramina separated by a thin horizontal bony bridge. The dorsal foramen, facing laterally, is designated as the foramen masticatorium, and the ventral one, facing ventrally, as the foramen ovale inferium. A suture is clearly visible in specimens IVPP V7151.61-63 and V7151.74 between the posterior margin of the part of the alisphenoid below the arbitrary line and the anterior lamina of the petrosal. The suture zone is immediately behind the posterior margins of the foramina masticatorium and ovale inferium. Thus both foramina are within the alisphenoid in lateral view. However, in medial (or endocranial) view, the posterior margin of the foramen masticatorium is bordered by the anterior lamina of the petrosal, due to the overlapping in the suture zone between the alisphenoid and the

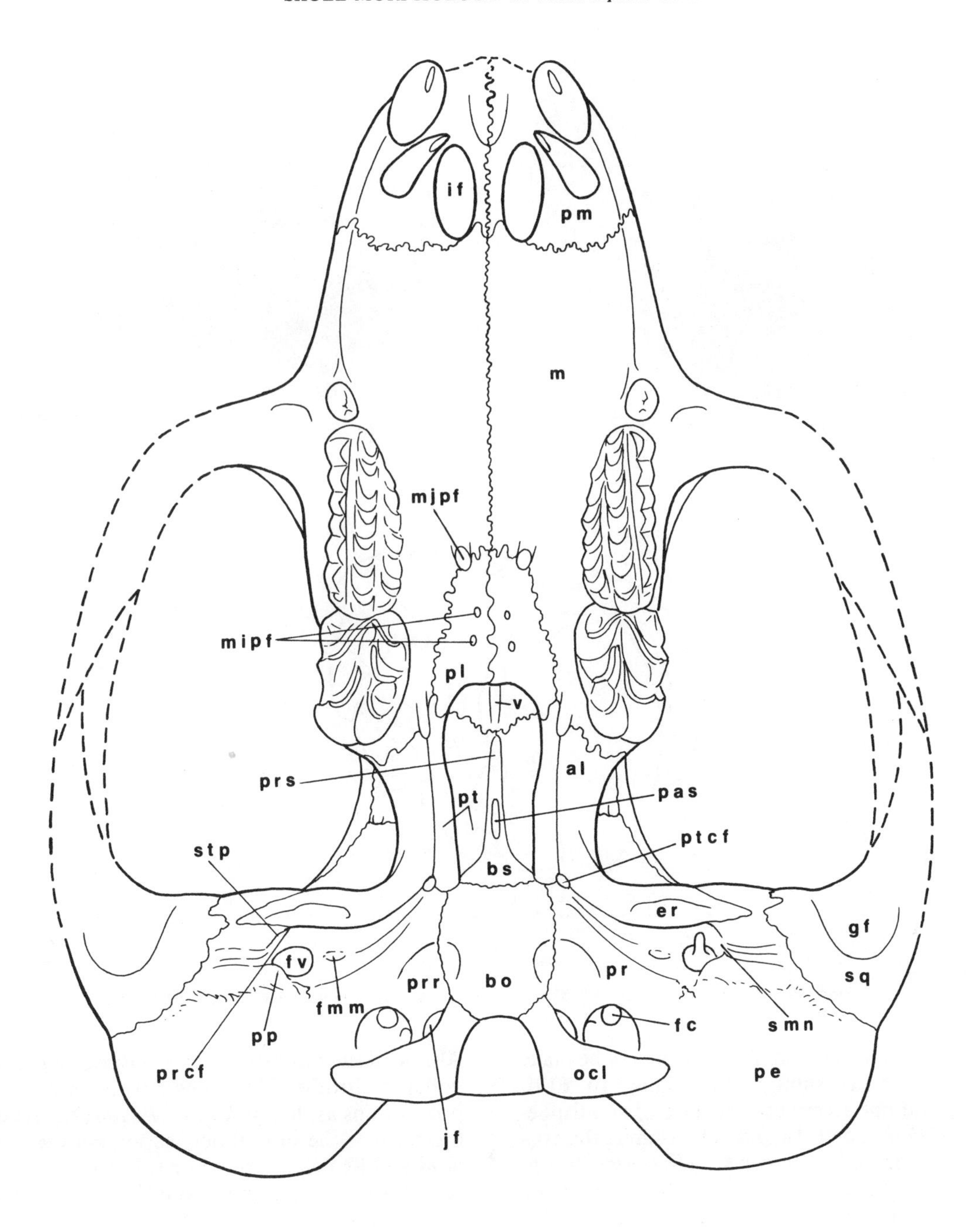

anterior lamina (that is, the alisphenoid laterally overlaps the anterior lamina of the petrosal).

The part of the alisphenoid above the arbitrary line is roughly fan-shaped, with its spread-out edge curved dorsally. It is in sutural contact with the descending process of the frontal anteriorly, with the anterior lamina of the petrosal posteriorly, and with the parietal dorsally.

Except for the anteroventral free edge (which laterally borders the sphenorbital fissure), the alisphenoid has a tight, sutural contact with all of its neighbors. In all cases, the alisphenoid overlaps the bones with which it makes contact.

Discussion

Alisphenoid in multituberculates and other nontherian mammalian groups.—Although the alisphenoid's contribution to the side wall of the braincase in multituberculates (and, in fact, in so-called "nontherians" as a whole) has been discussed extensively (*e. g.,* Kermack and Kielan-Jaworoska, 1971; Presley and Steel, 1976; Clemens and Kielan-Jaworowska, 1979; and Kielan-Jaworowska and others, 1986), it is surprising how few data on the alisphenoid are actually available.

Gidley (1909) mentioned possible presence of the alisphenoid canal and alisphenoid bulla in *Ptilodus*, but was unable to describe the alisphenoid bone itself because

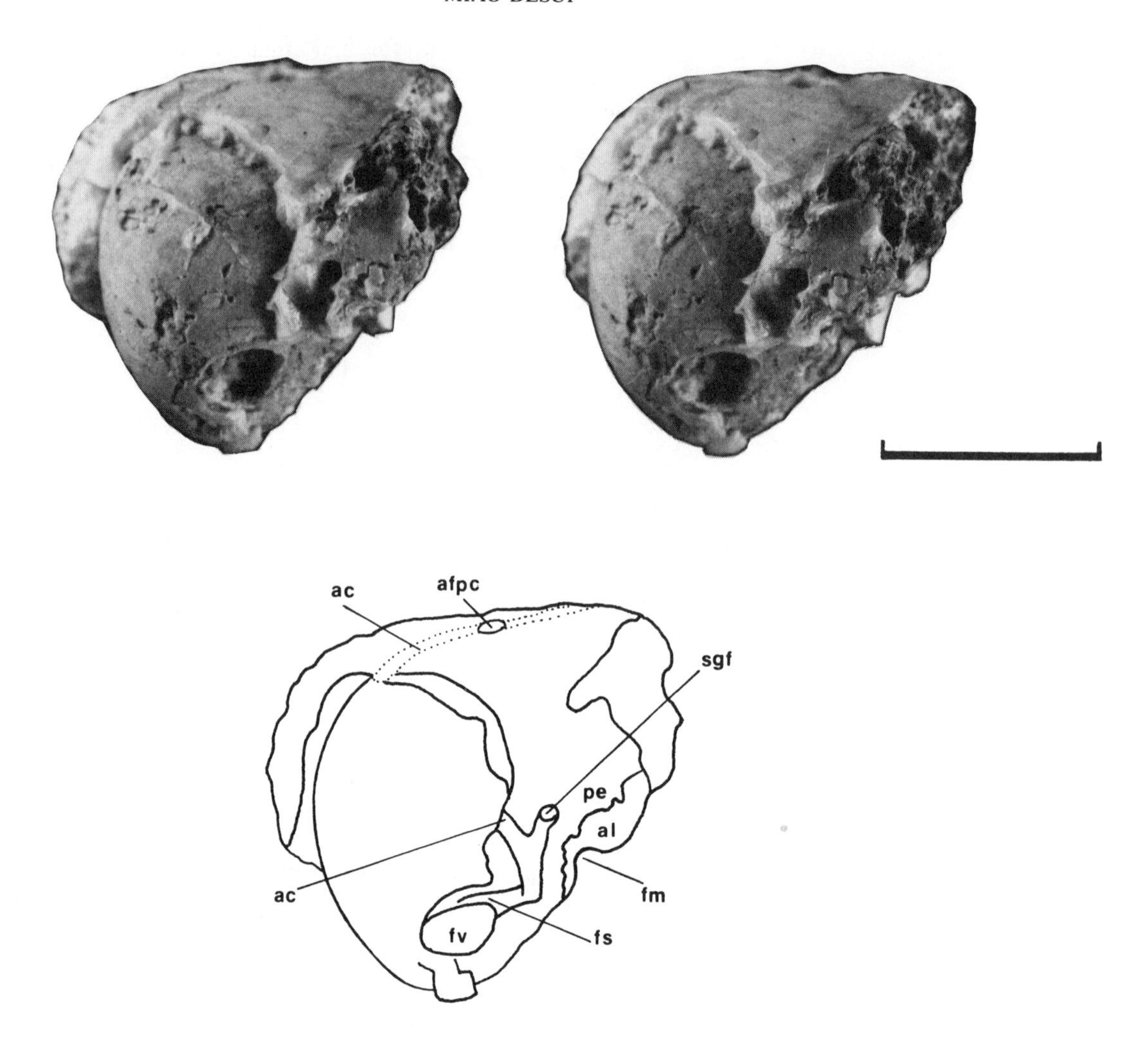

Figure 19. ***Lambdopsalis bulla*, V7151.61, anterolateral stereoscopic view of right petrosal.**

"The characters of the basicranial region can not be made out clearly, owing to crushing and breaking" (p. 619). Broom (1914) did not accept the presence of an alisphenoid canal in *Ptilodus*, and also did not recognize the existence of an alisphenoid bone per se. However, Broom indicated the presence of ". . . two large foramina in the alisphenoid region" (p. 123), and showed them in his Figure 3. Subsequently, Simpson (1937) suggested the absence of both the alisphenoid canal and bulla in *Ptilodus*, and interpreted one of the two foramina as the foramen ovale inferium and the other as an artifact. Again, as to recognizing the alisphenoid and its neighbors, Simpson was unable to "make out any sutures with sufficient probability to warrant their description" (p. 747).

Kielan-Jaworowska (1971) reconstructed the alisphenoid in *Kamptobaatar* as "a small, ventral element which does not contribute to the lateral wall of the braincase" (p. 12). However, she (*ibid.*, p. 12) admitted: "The recognition of the limits of the alisphenoid in the studied skull is also to some extent tentative, as the bone is cracked in this region and I cannot be sure whether I am dealing with sutures or with cracks on the bone surface." Although she reconstructed the alisphenoid in other Mongolian multituberculate genera with essentially the same proportions as that of *Kamptobaatar*, she repeatly made it clear that the sutural delineation between the alisphenoid and its neighboring bones is not recognizable with certainty (also see Clemens and Kielan-Jaworowska, 1979 and Kielan-Jaworowska and others, 1986, for further discussion). None of the cranial materials from Guimarota has been reported to have a recognizable alisphenoid (Hahn, 1969, 1977, and 1981). Sloan (1979) illustrated an alisphenoid in *Ectypodus* with its shape and proportion quite reminiscent of the alisphenoid generally seen in modern therians. However, lack of description of the bone in *Ectypodus* leaves open the question of how well the sutural demarcation of the alisphenoid can be observed in the actual specimen. As illustrated in Sloan's figure, at least the anterodorsal limit of the alisphenoid was reconstructed with a dashed line.

To the best of my knowledge, the above cases represent the entirety of published data from which the historically single most important character in phylogenetic analysis of early mammalian relationships (that is, the size of the alisphenoid) has been derived. A reduced alisphenoid seems to be extremely uncertain in multituberculates. Nevertheless, it has been used to argue that "Multituberculata and Monotremata are clearly related" (Kermack and Kielan-Jaworowska, 1971, p. 103). The same character also has been overemphasized in phylogenetic analyses to argue for a nontherian affinity of multituberculates (*e. g.,* Kermack and Kielan-Jaworowska, 1971; Kermack and others, 1981; and Kermack and Kermack, 1984), or for a therian affinity of monotremes (Kemp, 1983).

Monotremes are the only living representatives of "nontherian" mammals and, indeed, we know that they possess a small alisphenoid. Within extinct "nontherians," however, the only group in which the alisphenoid presumably has been well described is in morganucodontids. But in all known cases, morganucodontids have a relatively large alisphenoid with an extensive ascending process (see Kermack and others, 1981, for *Morganucodon*; Crompton and Sun, 1985, for *Eozostrodon heikuopengensis*; and Gow, 1986, for *Megazostrodon*). Nevertheless, Kermack and Kermack (1984, p. 55) claimed: "Although *Morganucodon* differs radically from the living Theria in these particulars, it has similarities with living monotremes (Kermack and Kielan-Jaworowska, 1971). The latter have a large anterior lamina and tiny alisphenoid and the sharing of this specialised character gives the most powerful reason for placing both *Morganucodon* and the monotremes in the sub-class Atheria."

As described above, the alisphenoid in *Lambdopsalis* is an extensive element in the orbitotemporal region, proportionately more extensive than that of morganucodontids. Even though such a feature may not characterize *all* of the multituberculates, it is instructive to contrast the positive evidence found in *Lambdopsalis* with suppositions as reviewed above.

Furthermore, even in *Morganucodon*, the confines of the alisphenoid as drawn by Kermack and others (1981) are questionable. They described (p. 115): "Behind the alisphenoid, the lateral braincase wall is formed by the anterior lamina of the petrosal (ant. lam. pet.). The ascending process of the alisphenoid lies lateral to the anterior lamina, there being a definite gap between the two bones." Subsequently, Kermack and Kermack (1984) even called the space between the two bones the cavum epiptericum. However, Hopson (*personal communication*) considers the gap between the alisphenoid and anterior lamina actually to be an artifact. My personal observation of the specimen confirmed Hopson's suggestion. In addition, both Crompton and Sun (1985) and Gow (1986) pointed out that there is a tight junction between the alisphenoid and anterior lamina in *Eozostrodon heikuopengensis* and in *Megazostrodon*. It seems clear that in *Morganucodon* we simply are not sure where the alisphenoid stops and the anterior lamina begins.

No generalizations about comparative extent of the alisphenoid can be made between multituberculates and other so-called nontherian fossil mammals.

Considerable confusion also has arisen from reconstructions by Kermack and his associates (Kermack, 1963; and Kermack and others, 1981) of topographic relationships between the alisphenoid and anterior lamina in *Morganucodon* and some other triconodonts. That is, they claimed that the alisphenoid lies lateral to the anterior lamina in these taxa. For example, Presley (1980, p. 161) was puzzled: "If the anterior lamina of the petrosal of triconodonts lies in a more medial plane with respect to the epipterygoid [see Kermack, 1963; Hopson, 1964, for discussion] then the triconodonts cannot be grouped with monotremes on this feature and the monotremes and cynodonts remain more like ditremes" [in that the anterior lamina meets the alisphenoid edge-to-edge]. However, as mentioned earlier, what Kermack and his associates called "anterior lamina of the petrosal" in these triconodonts is not equivalent to the anterior lamina in monotremes and cynodonts. It actually is the ossified pila antotica, which forms the medial wall of the cavum epiptericum and, indeed, is medial to the epipterygoid (or alisphenoid). In contrast, the "anterior lamina" in *Morganucodon*, as identified by Kermack and others (1981), accords with conventional usage of the term. Nevertheless, my personal observation of the specimen of *Morganucodon* referred to by Kermack and others (1981) suggests that the anterior lamina and the alisphenoid are in the same plane (contrary to their description of the former being medial to the latter).

Additionally, in his argument against Kemp's (1983) hypothesis about the sister group relationship between tritylodonts and mammals, Sues (1985) pointed out that whereas the anterior lamina laterally overlaps the ascending process of the epipterygoid in *Bienotherium* (see Hopson, 1964), the lamina is medial to the alisphenoid in *Morganucodon* (see Kermack and others, 1981). Therefore, Sues (1985, p. 208) argued: "These different topographical relationships suggest that these structures in the Tritylodontidae and in *Morganucodon* are independently derived." Although Sues' reasoning at first reading seems logical, as mentioned above, actual topographic relationships in the suture zone between the anterior lamina and the alisphenoid in *Morganucodon* are yet unknown.

Furthermore, it should be noted that the lateral overlapping of the epipterygoid in *Bienotherium* by the anterior lamina was considered by Hopson (1964) as strong evidence for the anterior lamina being an intramembranous (rather than neurocranial) ossification. Presley and Steel (1976) also suggested that the bony lamina (ossified from the outer surface of the spheno-obturator membrane in monotremes) lies lateral to both the periotic and alisphenoid. By accepting these premises, it is readily understandable why in *Bienotherium* and in monotremes (both with an extensive anterior lamina) the anterior lamina laterally overlaps the alisphenoid whereas in *Lambdopsalis* and in modern therians (both with an expanded alisphenoid) the alisphenoid laterally overlaps the anterior lamina or petrosal. This change in topographic relationship between the two bones can be accounted for by the change in the affinity of synostosis of the lamina obturans with either ala temporalis or petrosal. In other words, if the lamina obturans fuses with the ala temporalis and becomes part of the alisphenoid (as presumably is the case in adult *Lambdopsalis* and therians), it is the suture between the posterior margin of the lamina obturans and the petrosal that marks the alisphenoid/petrosal boundary. Therefore, it would be expected that the alisphenoid laterally overlaps the petrosal. If, on the other hand, the lamina obturans fuses with the petrosal and becomes the anterior lamina, it is the suture between the anterior margin of the lamina obturans and the alisphenoid that would mark the border between the alisphenoid and anterior lamina. Thus, it would be expected that the anterior lamina laterally overlaps the alisphenoid, as seen in *Bienotherium* and in monotremes.

Conversely, the topographical relationships in the edge-to-edge sutural contact between the alisphenoid and the anterior lamina (or petrosal itself) in advanced cynodonts and mammals can be used deliberately to test Presley's (1981) hypothesis of the presence during development of an independent membrane-bone between the ala temporalis and periotic. Presley's hypothesis would be seriously challenged if it were shown that: (1) a mammal (fossil or living) with an extensive anterior lamina in front of the petrosal has the lamina laterally overlapped by the alisphenoid; or, contrariwise, (2) a mammal with an expanded alisphenoid has its alisphenoid laterally overlapped by the petrosal. However, I am not aware of examples of either of these hypothetical cases. Thus Presley's (1981) interpretation of the side wall of mammalian braincase seems to be consistent with available evidence (see below for further discussion).

Foramina masticatorium and ovale inferium.—Two foramina in the expected location of the foramen ovale have been described in all multituberculate skulls in which that region is preserved (see Simpson, 1937; Kielan-Jaworowska, 1971; Sloan, 1979; and Kielan-Jaworowska and others, 1986). However, various authors have interpreted functions and the identity of surrounding bone(s) of the two foramina differently. Consequently, the foramina have various names among different taxa.

Simpson (1937) argued convincingly that both foramina in *Ptilodus* were for passage of the mandibular branches of the trigeminal nerve, based upon comparisons with modern rodents. He named the anterodorsal foramen the foramen masticatorium (for the exit of the masticatory nerve) and the posteroventral one the foramen ovale inferius (for the passage of the inferior ramus of the mandibular branch). Because of fusion between the alisphenoid and petrosal in available specimens of *Ptilodus*, however, Simpson was not able to specify which bone(s) these foramina perforate. Incidentally, in his description of the bony ear region of monotremes, Simpson (1938), by following Gaupp (1908), made a supposed distinction between the foramen ovale enclosed in the alisphenoid (in therians) and the foramen pseudovale surrounded by the anterior lamina of the petrosal (presumably in monotremes, but see discussion below). However, Simpson (1937) did not apply the term foramen pseudovale to multituberculates. Unfortunately, this distinction, though trivial as originally anticipated, has been greatly exaggerated, often without concrete evidence.

Kielan-Jaworowska (1971) described two foramina in the ventral part of the orbitotemporal region in *Sloanbaatar* and five in *Kamptobaatar*. However, she considered all of them as passages for branches of the mandibular nerve (following Simpson). Kielan-Jaworowska interpreted one foramen in *Sloanbaatar* (and four in *Kamptobaatar*) as perforating the anterior lamina; the remaining foramen was deemed to pass between the anterior lamina and the alisphenoid. While it was impossible to ascertain which foramina gave passage to which branch of the mandibular nerve in *Kamptobaatar*, she suggested that in *Sloanbaatar* the more anterior and ventral one transmitted the inferior ramus of the mandibular nerve. The more posterior and dorsal foramen was interpreted to transmit the masticatory ramus. Although she considered the foramen for the inferior ramus to lie between the alisphenoid and the anterior lamina and the foramen for the masticatory ramus as being within the anterior lamina, she nevertheless called the former the foramen pseudovale inferius and the latter the foramen masticatorium. Subsequently, both of these two foramina in the other Mongolian multituberculates were interpreted as perforating the anterior lamina, and the prefix "pseudo-" used in the term "foramen pseudovale inferium" was abandoned (Kielan-Jaworowska and others, 1986).

Sloan (1979) illustrated both foramina in *Ectypodus* as passing through the alisphenoid, and suggested that they served for passage of the maxillary and mandibular branches of the trigeminal nerve. Although neither interpretation of functions of the foramina can be fully justified, their closer topographic association as seen in *Lambdopsalis* and other multituberculates would seem to strengthen Simpson's (1937) analysis. As described above, the two foramina in *Lambdopsalis* are separated by only a slender bony bar, which is vulnerable to loss during preparation. It also should be noted that although relative positions of the two foramina in multituberculates may vary slightly, they are always close and separated only by a tiny bony bridge. In addition, Simpson's (1937) functional consideration of the parallel developments of a double orifice for the mandibular nerve in multituberculates and rodents remains attractive. It seems reasonable to accept Simpson's interpretation and terminology for the two foramina in absence of evidence to the contrary.

Closely related to the problem of sutural delineation (and thus extent) of the alisphenoid as discussed earlier, bones surrounding the foramen ovale also have been used in attempts to decipher mammalian phylogeny. Kermack and his associates (Kermack and Mussett, 1958; Kermack, 1963, 1967; Kermack and others, 1981; and Kermack and Kermack, 1984) interpreted the foramina for passages of the maxillary and mandibular branches as lying within the extensive anterior lamina of the petrosal in morganucodontids and other triconodonts, as in modern monotremes. Contrariwise, these foramina in therians were presumed to perforate the expanded alisphenoid.

Accordingly, they followed Gaupp's (1908) terminology, but overemphasized the distinction between the foramina pseudovale and ovale by further suggesting that they are characteristic of nontherian and therian mammals, respectively.

Up to the late 1970s, this view dominated the mainstream of thinking among paleomammalogists. However, flaws in the basis upon which Kermack's hypothesis was built have at least three sources. First, as to the foramen pseudovale, MacIntyre (1967, p. 834) stated: ". . . no evidence has yet been published which demonstrates clearly that such a relationship between the mandibular nerve and the periotic bone really existed in morganucodonts (or, indeed, in any known reptile or mammal, living or fossil)." Two decades have passed and much of pertinence has been published since MacIntyre wrote this. It is now suggestive that, at least in *Morganucodon oehleri*, the foramen ovale perforates the anterior lamina of the petrosal (Hopson, *personal communication*). But MacIntyre's comment remains valid as far as the other "nontherian" taxa are concerned.

As discussed above, prior to the present study there has been no unequivocal evidence showing the exact boundary between the alisphenoid and the anterior lamina in any early mammal. That observation alone should throw doubt on interpretations of the bone surrounding the foramen ovale as being the petrosal in all so-called "nontherian" mammals. Besides, even in monotremes (usually considered as the only living representatives of the paraphyletic nontherians), it has been demonstrated that the foramen for the mandibular branch of the trigeminal nerve is situated between the alisphenoid and the anterior lamina in *Ornithorhynchus*, and between the pterygoid and alisphenoid in *Tachyglossus* (see Watson, 1916; MacIntyre, 1967; and Griffiths, 1968, 1978). In neither case does the mandibular nerve perforate the petrosal, contrary to Kermack's claim.

It is interesting that, despite his strong disagreement with Kermack, MacIntyre (1967) regarded the term foramen pseudovale descriptively meaningful for the opening lying between the alisphenoid and petrosal (presumably for passage of the mandibular nerve). This usage was followed by Archer (1976), but rejected by Griffiths (1978) and Kielan-Jaworowska and others (1986).

Moreover, although it was impossible to determine which bone the mandibular nerve perforated in other multituberculates, the foramina masticatorium and ovale inferium in *Lambdopsalis* as described herein clearly show that they are mainly enclosed by the alisphenoid, with only minor posterior marginations contributed by the petrosal (and only in medial view).

Also, the bone posteroventral to the trigeminal foramina, which includes what Kielan-Jaworowska and others (1986, Fig. 6) called "ridge of anterior lamina," "epitympanic recess," and "lateral flange" in the Mongolian taxa, is actually part of the posterior process of the alisphenoid in *Lambdopsalis*. This is also what Miao and Lillegraven (1986, p. 502) called the "bony bridge that abuts medially against the expected position of the alisphenoid and laterally against the expected position of the petrosal." Lacking a recognizable suture in that specimen, Miao and Lillegraven (1986, Fig. 1) denoted it as "?petrosal or alisphenoid." Subsequently, I have found a clear sutural delineation of the bone in IVPP V7151.94, demonstrating it as part of the alisphenoid (comparable to the quadrate ramus of the epipterygoid or alisphenoid in cynodonts and *Morganucodon*). Therefore, at least in *Lambdopsalis*, the ventral part of the petrosal does not have an anterior lamina at all.

The second problem that exists with Kermack's hypothesis is that, contrary to claims of Kermack and his associates, the foramen ovale in therians does not universally pass through the alisphenoid. In fact, position of the foramen ovale is highly diverse in therians, not only among orders, but also within genera. For example, Archer (1976) documented that in some marsupicarnivores the foramen for the mandibular nerve ". . . is formed antero-laterally by a notch in the postero-medial corner of the alisphenoid and postero-medially by the anterior tip of the petrosal" (p. 229). He even chose to call it the foramen pseudovale, following MacIntyre (1967), because the foramen does not perforate the alisphenoid alone. In addition, Gregory and Noble (1924) mentioned that the foramen ovale in *Perameles obesula* is located at the anterior edge of the alisphenoid. Griffiths (1978, p. 325) added ". . . one other variation on the theme—in the marsupial *Vombatus ursinus* the anterior half of the foramen ovale is bordered by the alisphenoid, the posterior half by the squamosal."

Although the foramen ovale is enclosed within the alisphenoid in most eutherians, Edinger and Kitts (1954) recorded variation in its position within the order Perissodactyla from confluence with the foramen lacerum medium to a location between the alisphenoid and petrosal.

Finally, it has long been known that in development of mammals there is a spheno-obturator membrane between the ala temporalis and the otic capsule, from which an intramembranous bone develops in later stages (Gaupp, 1908; Watson, 1916; Gregory and Noble, 1924; Muller, 1935; De Beer, 1937; Hopson, 1964; Vandebroek, 1964; MacIntyre, 1967; Griffiths, 1968, 1978; Presley and Steel, 1976; and Presley, 1981). It also has long been known that the mandibular branch of the trigeminal nerve leaves the cranial cavity through that intramembranous bone, or through the space between the membrane bone and the petrosal (Watson, 1916; De Beer, 1937; and Presley and Steel, 1976). As mentioned earlier, Presley (1981) considered trivial the differences in extent of the alisphenoid and in the position of the foramen ovale between the nontherian and therian mammals by postulating a minor change of synostosis of that intramembranous bone with its neighbors— the alisphenoid or petrosal. In addition, Presley (1981, p. 670) claimed: "It must be appreciated that the blade of the alisphenoid and the anterior process of the petrosal are developmentally very similar and may be morphologically almost equivalent." If such is indeed the case, the foramen for the passage of the mandibular nerve appears always to be surrounded completely (or mainly) by that morphologically equivalent membrane bone. The resulting position for the foramen thus solely depends upon with which bone the morphological equivalent fuses in development. Therefore, the presumably fundamental difference in the bony surroundings of the foramen ovale between nontherians and therians seems more apparent than real (even if such a difference indeed exists).

This view was essentially endorsed more recently by Kuhn and Zeller (1987). They pointed out that, developmentally in most therians (except for cases such as *Erinaceus* and *Homo*), the mandibular branch of the trigeminal nerve leaves the cavum epiptericum primarily posterior to (rather than within) the ala temporalis, the cartilaginous anlage of the alisphenoid. The final enclosure of the foramen ovale into the alisphenoid is formed secondarily by the periosteal ossification of the ala temporalis later in ontogeny. This again implies that no substantial difference in phylogeny need be invoked even if the presumed difference in bony surroundings of the foramen ovale between nontherians and therians indeed existed.

More than three decades ago, Edinger and Kitts (1954, p. 391) wrote: "While zoologists decided that the foramen ovale is a feature of great taxonomic importance in ungulates and of evolutionary significance in mammals as a whole, paleontologists paid little attention to that small detail." Ironically, right after their comment was published, paleontologists started to reverse that tradition.

Evolution of mammalian alisphenoid.—As Struthers (1927, p. 179) noted: "The phylogenetic history of the alisphenoid of mammals presents one of the most perplexing and at the same time basic problems centering around the orbitotemporal region." Vast literature, most appearing in the first quarter of the Twentieth Century, in developmental studies of skulls of various mammalian taxa contains extensive discussions of the patterns and processes of development of the mammalian alisphenoid (*e. g.*, Gaupp, 1900, 1902; Broom, 1907, 1909; Fuchs, 1915; Watson, 1916; Terry, 1917; Gregory and Noble, 1924; Struthers, 1927; Goodrich, 1930; and Muller, 1935). Controversies around the problem of origin of the mammalian alisphenoid were summarized lucidly by De Beer (1937).

It seems that recent revival of these otherwise-forgotten controversies has been primarily because some paleontologists are unaware of them. In the past two or three decades, for example, paleontologists started to pay increasingly close attention to the enigmatic alisphenoid bone. They claimed that a fundamental distinction exists between nontherian and therian mammals in the extent of the alisphenoid's contribution to formation of the side wall of the braincase. Whereas monotremes and their presumed nontherian relatives are said to have an insignificant contribution from the alisphenoid to enclose their expanding braincase, therians are thought to be characterized by an expansion of the alisphenoid to form ample parts of the lateral wall of the braincase. The presumed reduction of the alisphenoid has been deemed a unique specialization of the braincase for all nontherian mammals (Kermack, 1967; Hopson, 1969, 1970; Hopson and Crompton, 1969; Crompton and Jenkins, 1979; Kermack and others, 1973, 1981; Kermack and Kielan-Jaworowska, 1971; and Kermack and Kermack, 1984).

Although these authors agreed (on the basis of the alisphenoid plus a presumably basic difference in dentition) upon an early separation of two major mammalian groups, they differed on whether early mammals were derived once, or more than once, from reptilian ancestors. Kermack and his associates conceived that the epipterygoid of cynodonts was too expanded to have given rise to the morganucodontid alisphenoid. Therefore, they reasoned that ancestry of morganucodonts (as well as other nontherians) must have been within another line of therapsid evolution, specifically the bauriamorphs, in which an increase in breadth of the ascending process of the epipterygoid did not occur. In contrast, Hopson, Crompton, and Jenkins held that the structure of the side wall of the cynodont braincase appeared to provide a starting point from which either the therian or nontherian pattern could have been derived. That is: (1) the therian pattern could have been obtained by a posterior expansion of the epipterygoid around the trigeminal nerve, coupled with reduction of the anterior lamina of the petrosal; and (2) the nontherian condition could have resulted from an anterior extension of the anterior lamina, together with reduction of the epipterygoid. However, these authors did not relate developmental processes involving the alisphenoid and their important implications to the discussion.

Meanwhile, however, Vandebroek (1964), MacIntyre (1967), and Griffiths (1968) pointed out that so-called anterior lamina of the petrosal in monotremes is an independent intramembranous ossification filled by an invasion of bone from the ala temporalis, and has nothing to do with development of the petrosal. Griffiths (1968) went further, and argued that the anterior lamina ". . . has no adaptive significance and therefore it seems unlikely to have arisen independently in *Morganucodon* and in monotremes" (p. 221). In other words, Griffiths suggested that, as in monotremes, the anterior lamina in *Morganucodon* also has affinity with the alisphenoid rather than the petrosal. However, these ideas were either ignored or rejected by other principals in the discussion. By the time of the London Symposium on early mammals held in 1970, the consensus of a fundamental dichotomy within the Mammalia was firmly established among most paleontologists in the field (Kermack and Kielan-Jaworowska, 1971), and those voices of dissent against the hypothesis of mammalian dichotomy were virtually ignored.

It is interesting to note that Simpson, as a doyen of the field, was uncommitted on the issue. He (1971) was impressed by the claimed differences in braincase structure between nontherians and therians, but he simultaneously cautioned that ". . . one should not cavalierly discard the suggestion of MacIntyre (1967) that the morganucodontid (or 'nontherian') condition may be primitive for mammals" (p. 191).

Traditionally, the paleontologists' concept of homology of the mammalian alisphenoid was the simplistic notion "that the mammalian alisphenoid has been derived from the cynodont epipterygoid" (Gregory and Noble, 1924, p. 459). However, as mentioned earlier, it has been generally accepted among embryologists that the mammalian alisphenoid is a composite bone, consisting of a cartilaginous ala temporalis (homologue of the epipterygoid) and a membranous lamina obturans (James and others, 1980). It is primarily through Presley's effort (Presley and Steel, 1976; and Presley, 1981) that these aspects of the development of the mammalian alisphenoid have been stressed, and thereafter have radically, but gradually, altered our viewpoint.

Presley and Steel (1976) examined developmental stages in the orbitotemporal regions of echidna, opossum, fruit bat, rat, rabbit, pig, and man, and confirmed that: (1) "The alisphenoid in mammals is part cartilage bone, part membrane bone" (p. 458); and (2) the element of membrane bone in Ditremata ". . . originated as the anterior lamina of the periotic in cynodonts, which is retained in monotremes" (p. 459). Subsequently, Griffiths (1978) described that a substantial intrusion of bone from the ala temporalis region was seen invading the membrane in the sphenoparietal fissure in skulls of young echidnas. He then considered this intrusion to form part of the alisphenoid, but not to connect it with the periotic. Griffiths, however, did not doubt the validity of Watson's (1916) interpretation of the anterior lamina as an outgrowth of the periotic in platypus. Therefore, Griffiths (1978, p. 313) reasoned that ". . . it is more than passing interest that in the two living representatives of what may well have been Mesozoic mammals the side walls of the braincase are filled in by bone of fundamentally different origin."

Although Kermack did not seem to doubt Griffiths' observation, he considered "this to be a specialization of that animal, and not significant" (Kermack and others, 1981, p. 131). However, Presley (1981) soon confirmed Vandebroek's (1964) suggestion that in the platypus the so-called anterior lamina also is not derived from a laminar outgrowth of the petrosal (incidentally, Vandebroek's original suggestion did not draw widespread attention). Presley (1981, p. 669) pointed out: "The area of interest develops in the lateral wall of the cavum epiptericum, the cartilaginous ala temporalis in front and the cartilaginous otic capsule (future petrosal) behind. In all mammals a fibrous spheno-obturator membrane lies between these." Later in development a membrane bone, lamina obturans, is formed in the position of the spheno-obturator membrane. Presley concluded that there seems to be no fundamental difference between therians and nontherians in early development of the lamina obturans (which is virtually an independent membrane bone); the difference arises only when the lamina obturans fuses with its neighbors, either the ala temporalis (anteriorly) or otic capsule (posteriorly).

According to Presley (1981, p. 670), therefore, "If Mesozoic, like recent, mammals formed membrane bone in the area of the lamina obturans, it follows that any form in the fossil record with an expanded epipterygoid, an anterior process of the petrosal, or both, could, by a simple change in the affinity of synostosis during development, come to possess therian anatomy." Thus the nature of the braincase structure in morganucodontids, triconodonts, multituberculates, docodonts, or even monotremes provides little useful phylogenetic information; certainly it does not exclude their close affinity to therians, and does not prove polyphyly or diphyly in the origin of mammals.

Presley's postulation seems plausible, and indeed has been proved to be commensurate with repeatable observation. The argument was developed primarily on embryological grounds. But, expressed explicitly by Presley and Steel (1976, p. 450), "By themselves, such grounds may be quite misleading." Discovery of a relatively large alisphenoid and narrow anterior lamina of the petrosal in *Lambdopsalis* provides strong support from the fossil record for Presley's argument. The discovery clearly demonstrates the variability of ossification of bony elements in the orbitotemporal region (as allowed by developmental processes), and the danger of employing morphology of this region for rigid phylogenetic interpretations. This, in my opinion, embraces the essence of Presley's argument, in contrast to some other representations (*e. g.*, Kemp, 1983) that were actually contrary to Presley's intentions.

However, this new piece of supporting evidence was not available to Presley and Steel (1976). Although they were restricted by the scarcity of well preserved cranial materials, they nevertheless managed to find support for their hypothesis from the position of the trigeminal foramina as seen in cynodonts. By following Parrington (1946), Hopson (1964), and Hopson and Crompton (1969), Presley and Steel (1976) accepted that the trigeminal foramen lying in the suture line between the epipterygoid and the anterior lamina in cynodonts was the exit for both the maxillary and mandibular branches of the trigeminal nerve. They also cited in what Parrington (1946) described as "*Trirachodon*" (a specimen of *Scalenodon angustifrons*) that two trigeminal foramina exist as separated exits for the maxillary and mandibular branches. These two foramina are separated by a bony bar posterior to the true processus ascendens of the epipterygoid. Presley and Steel regarded these foramina as strong evidence indicating that: (1) cynodonts retained ". . . the fundamental relationship of those nerves to the processus ascendens found throughout the vertebrates except, apparently, in recent mammals" (Presley and Steel, 1976, p. 452; that is, both the nerve branches are posterior to the processus ascendens); and (2) the bony bar separating the two trigeminal foramina in "*Trirachodon*", dubbed as the lamina ascendens, represents a neomorphic structure, and is not homologous with the processus ascendens.

These interpretations seem to aid understanding of the transition from a typical nonmammalian vertebrate condition to the mammalian condition, simply by assuming existence of a trend toward reduction of the processus ascendens and increasing upgrowth of the lamina ascendens. Apparently, this view has its root in De Beer's (1926) idea, and inevitably has revived the old controversies over development of the mammalian alisphenoid.

Gaupp (1900, 1902) found that the processus ascendens of the epipterygoid in nonmammalian vertebrates lies between the profundus and maxillary branches of the trigeminal nerve, but the ala temporalis in mammals lies between the maxillary and mandibular branches. This led Gaupp to believe that the processus ascendens of the reptilian epipterygoid is not homologous with the mammalian ala temporalis, and he considered the former as probably lost and replaced by the latter in mammals. Broom (1911), however, did not consider the difference so significant, and stated: "In the mammal the alisphenoid differs from that of the cynodont mainly in having 2nd and 3rd branches of Nerve V. passing through instead of behind it" (p. 912). Broom held the view that the mammalian ala temporalis can be derived readily from the processus ascendens of the reptilian epipterygoid. However, Fuchs (1915) showed that in *Didelphis* the ala temporalis is situated between the profundus and maxillary branches, much as in reptiles. De Beer (1926, 1937) further pointed out in many mammals the cartilage exists both between the profundus and maxillary and between the maxillary and mandibular branches. He regarded the cartilage between the profundus and maxillary branches as the processus ascendens, and the one between the maxillary and mandibular branches as the ala temporalis (*i. e.*, Presley's "lamina ascendens"). In De Beer's view, therefore, the original cartilaginous part of the alisphenoid includes both the processus ascendens and lamina ascendens as well as the processus alaris (*i. e.*, the basitrabecular process). This has been essentially incorporated within Presley's argument.

Maier (1987), in his thorough review of the problem of the mammalian alisphenoid, revived Fuchs' (1915) idea. Maier further confirmed that not only in *Didelphis* but also in many other genera of didelphid marsupials, the processus ascendens of the ala temporalis is situated between the profundus and maxillary branches of the trigeminal nerve, as in reptiles. A lamina ascendens that Presley reported was not found. However, "The mode of ossification of the alisphenoid apparently varies considerably among these taxa" (Maier, 1987, p. 85). Maier also pointed out that variation in the relationship between the processus ascendens and branches of trigeminal nerve occurs even within the Dasyuridae. For example, Maier found that in a neonate *Sminthopsis rufigenis*, the processus ascendens is perforated by the maxillary nerve. In contrast, Broom (1909) showed that the processus ascendens lies between the maxillary and mandibular branches in a newborn *Dasyurus viverrinus*, much as in reported cases in most other mammals.

This sort of diversity also occurs in eutherians. Zeller (1986), for example, reported that in *Tupaia* the processus ascendens of the ala temporalis is perforated by the maxillary nerve. He considered that "This is an autapomorphous feature of *Tupaia*" (p. 276). In contrast, Reinbach (1952*a,b*) warned against attaching too much weight to the neural perforation of the ala temporalis as a reliable character of a group, and called for further research on the embryological origin and possible variations of the character. Reinbach was well aware that frequently only relatively late stages of the embryonic development have been examined. In addition, lack of appreciation of diversity in relationship between branches of the trigeminal nerve and the processus ascendens may be due to limited number of mammalian taxa that have been studied. According to Kohncke (1985), the chondrocranium has been studied in less than 60 of the roughly 1,000 currently recognized mammalian genera (Anderson and Jones, 1984). Thus, McKenna's (1976) jest remains valid that the typologists' monotonous litany of "the" mammals sounds like a one-note concerto. Maier's (1987) and Zeller's (1986) discoveries offer testimonies to Reinbach's (1956a, b) concern.

Furthermore, Maier (1987) postulated a fuctional reason for inevitability of such diversity. He pointed out: ". . . in early embryogenesis the peripheral nerves are much earlier developed than the first skeletal structures; the latter have to adapt to the primary organization of the other head organs" (p. 85). As such, Rajtova's (1972, p. 190) description of both the first and second trigeminal branches being posterior to the ala temporalis in *Cavia* are credible (see also Kohncke, 1985). It is also conceivable that ". . . differential embryonic brain growth may have shifted the position of the ascending process and the trigeminal nerves. Therefore, it seems quite unnecessary to make a formal distinction between an anteriorly situated ascending process (being homologous to the epipterygoid) and a posteriorly lying Lamina ascendens (corresponding to a neomorphic mammalian alisphenoid) as suggested by De Beer (1926; 1937) and Presley and Steel (1976)" (Maier, 1987, p. 86). To do otherwise (as Presley and Steel, 1976, suggested), one must consider the reemergence of the processus ascendens in didelphids and *Tupaia* as evolutionary reversals; although possible, it seems less parsimonious.

Paleontological evidence cited by Presley and Steel (1976) was challenged by Sues (1986). According to Sues, what Presley and Steel regarded as the precursor of the mammalian ala temporalis (the bony partition between the two trigeminal foramina in "*Trirachodon*") is nonexistent. Sues found no indication of such a process in specimens of his newly-described *Kayentatherium*, which also possesses separated foramina as the exits for the maxillary and mandibular branches. Sues (1986, p. 237) stated: "in all these examples, the epipterygoid forms the anterior margins of the trigeminal foramina. The mammalian alisphenoid possibly includes the processus ascendens of nonmammalian synapsids *and* part of the anterior lamina of the periotic" (italics original). By speaking of "the anterior lamina of the periotic," Sues meant the intramembranous ossification—the lamina obturans of Presley (1981).

As a concluding remark on the alisphenoid's development in neonate marsupials, Maier (1987, p. 88) wrote: "The ascending process of the Ala temporalis then becomes strengthened by perichondral ossification and 'zuwachsknochen' (appositional bone), which spreads into the sphenoobturatory membrane; these ossifications are called the alisphenoid."

Clearly, all these authors agreed upon the concept that the mammalian alisphenoid includes both the reptilian epipterygoid and an appositional bone, the lamina obturans. This accords fully with Presley's main argument of the composite nature of the mammalian alisphenoid and its implications for understanding mammalian evolution. But some authors differ from Presley only in minor details of certain aspects of developmental processes. I applaud Maier's functional interpretation about the possible diversity in relationship between branches of the trigeminal nerve and the processus ascendens in mammals. However, I think that one cannot overemphasize the importance of Presley's effort in drawing paleontologists' attention to developmental aspects of the mammalian alisphenoid; as a consequence, many paleontologists' comprehension of early mammalian evolution has been changed. After all, Presley claimed originality only for his interpretation of the patterns of ossification of the mammalian alisphenoid as applied to mammalian phylogeny. He did not claim originality for descriptions of embryonic materials of living mammals (see also James and others, 1980).

Rudwick (1976), a philosopher and historian of science, once wrote: "It would be salutary for the historians to be reminded by the palaeontologists that most of the earlier frames of reference are still being utilised—even if unrecognised—in modern palaeontology: the insights and methods of one 'paradigm' of interpretation have not been wholly abandoned, but absorbed into the next" (p. 267). In the present context, paleontologists certainly have found companions among embryologists. In fact, any practitioner in any historically-rooted scientific discipline cannot avoid being influenced in one way or another by his intellectual heritage. Even the above-quoted Rudwick's comment can find its parallel in G. K. Gilbert's (1896, p. 2) writing: "Just as in the domain of matter nothing is created from nothing, just as in the domain of life there is no spontaneous generation, so in the domain of mind there are no ideas which do not owe their existence to antecedent ideas which stand in the relation of parent to child. It is only because our mental processes are largely conducted outside the field of consciousness that the lineage of ideas is difficult to trace." It is in this spirit that we should do justice to Presley's contributions to vertebrate paleontology.

Some criticism of Presley's work simply falls into the category of the critic's own misunderstanding. For example, Kuhn and Zeller (1987, p. 67) stated: ". . . the ascending lamina of the alisphenoid of Theria and the Lamina obturans of monotremes can not be regarded as morphologically 'equivalent' (PRESLEY, 1981) or homologous. In addition, it is very unlikely that the alisphenoid is derived from the Lamina obturans 'by a simple change in the affinity of synostosis during development' (PRESLEY, 1981, p. 670). KEMP (1982, 1983) seems to follow PRESLEY in essential positions. Such changes have never been documented for any skeletal element. In addition, topographic relations and ontogeny of the Lamina ascendens of therians and of the Lamina obturans of monotremes make it most unlikely that such a homology exists." No, of course not. The problem, however, is that Presley (1981) never made such a claim as the homology between the ascending lamina of the alisphenoid of Theria and the lamina obturans of monotremes. Quite contrary to this, Presley (1981) made his terminology crystal clear in the opening paragraph of his paper: "Here I use 'processus ascendens' for the part of the ala between ophthalmic and maxillary nerves; 'lamina ascendens' for the part of the ala between maxillary and mandibular nerves. . . I propose here that the more recent term 'lamina obturans' used for membrane bone developing in the spheno-obturator membrane of monotremes may usefully be extended to the very similar field of membrane bone here in therians" (p. 669). Obviously, Presley made the clear distinction between the part of the ala—"lamina ascendens" (being a cartilage replacement bone) and the "lamina obturans" (being membrane bone). By the very definition, Presley only equated the lamina obturans of monotremes with the lamina obturans (but not the lamina ascendens) of therians. As mentioned earlier in this section, Kemp (1982, 1983) also misinterpreted the essence of Presley's argument (Presley, *personal communication*). The only valid element in Kuhn and Zeller's criticism, in my opinion, is that indeed Presley did not explain why and how the changes in the affinity of synostosis between the bones took place.

Kuhn and Zeller (1987) also opposed Presley and Steel's (1976) notion of homologies between the lateral walls of the cavum epiptericum in monotremes and therians, primarily by claiming the differences in the detailed ossification patterns of the lateral walls of the cavum. They further suggested that both the lateral walls of the cavum and the cavum epiptericum itself are independently developed in *Tachyglossus, Ornithorhynchus*, and therians. As discussed extensively earlier in this paper, the homology of the cavum epiptericum throughout vertebrates has been well documented and widely accepted. Kuhn and Zeller's view only represents a different value of the term "homology," or simply a choice of interpretation. For example, the forelimbs of pterosaurs, birds, and bats are said to be homologous in terms of their historical genesis, but analogous in terms of acquisitions of their current utilities.

In fact, the problem raised by Kuhn and Zeller was already well thought out by Presley (1980) in developing his hypothesis about the intramembranous ossification of the lateral wall of the cavum epiptericum. Presley (1980, p. 160) argued: "If the exact pattern of ossification on this membrane is to have much taxonomic weight, then *Ornithorhynchus* must be regarded as well-separated from *Tachyglossus*, perhaps with as much force as are cynodonts separated from mammals. It is more reasonable to regard this area of membrane bone as a region whose pattern in advanced therapsids is very labile." The fact that Kuhn and Zeller (1987) still considered *Tachyglossus* and *Ornithorhynchus* as a natural group (the monotremes) trivializes the taxonomic significance of diversity in the detailed patterns of ossification of the region. This seems exactly what Presley intended.

To sum up, the mammalian alisphenoid is derived from the reptilian epipterygoid plus an intramembranous ossification between the processus ascendens of the ala temporalis and the otic capsule. The alisphenoid is a homologous structure throughout the class Mammalia, and can be equated with the epipterygoid (or "alisphenoid") plus the anterior lamina of cynodonts, monotremes, and possibly some of the so-called "nontherian" mammals. The exact patterns of ossification in the area of the spheno-obturatory membrane vary among mammalian taxa. Therefore, the orbitotemporal morphology can be unreliable for purposes of rigid phylogenetic interpretation, contrary to traditional belief. Presley's main argument about implications of the developmental pattern of the mammalian orbitotemporal region remains valid and consistent with knowledge derived from embryology and comparative anatomy. Despite recent criticisms from fellow developmental biologists, Presley's hypothesis receives strong support from discovery of the large alisphenoid in *Lambdopsalis*. Nevertheless, the current debate cannot be settled until occurrence of some future: (1) paleontological discovery that could document a Mesozoic therian mammals with an anterior lamina; and/or (2) developmental study that could provide a satisfactory explanation for changes in the affinity of synostosis of the membranous ossification with its neighbors.

Petrosal and Auditory Region (Figs. 10, 12, 17-30)

Description

The petrosal and auditory region can be observed clearly in the following specimens: IVPP V5429, V5429.7, V5429.8, V7151.56-75, V7151.77, V7151.89, V7151.90, V7151.94, and V7151.95. In addition, there are many more nearly complete (but isolated) petrosals in the collection. Almost all necessary anatomical information about skeletal structures of the ear region can be pieced together from the specimens under description.

The petrosal is greatly expanded, and superficially bulla-like, occupying a disproportionately large part of the posterior one-fourth of the skull. The middle ear is completely open ventrally, and there is no indication of any bony elements which may have contributed to the formation of auditory bullae in life. Inside the expanded petrosal is a spheroidal, hollow space. The petrosal is in sutural contact with the parietal dorsomedially, with the alisphenoid anteromedially, with the squamosal anterolaterally, with the basisphenoid and basioccipital ventromedially, and with the supraoccipital and exoccipital posteromedially. Broken parts of the petrosal in various specimens reveal a highly cancellous structure inside the bone.

Occipital aspect of petrosal.—In occipital view, the petrosal is smooth-surfaced, greatly convex posteriorly, and teardrop-shaped with its tip pointing dorsomedially. The dorsal and lateral margins of the petrosal form a striking crest, here designated as the lambdoidal crest. The margins also mark the dorsal and lateral sutural contacts with the squamosal. The medial margin of the petrosal is in sutural contact with the supraoccipital and exoccipital, roughly in a straight descending line. In the ventromedial corner of its medial margin, there is a rough-surfaced small process just lateral to the occipital condyle and posterolateral to the fenestra cochlea. The occipital surface of the process is slightly concave.

Slightly ventral to, and at the mid-point of, the dorsal segment of the lambdoidal crest, there is a small fossa facing posterolaterally, here designated the post-temporal fossa. A groove extends laterally from the fossa, and fades out gradually at the location where it makes its descending turn (adjacent to the junction between the dorsal and lateral segments of the lambdoidal crest). A canal within the dorsal rim of the petrosal runs anterodorsally from the post-temporal fossa to join the ascending canal. As noted before, this canal is called the post-temporal canal. There is no direct connection between the post-temporal canal and the subarcuate fossa.

Anterolateral aspect of petrosal.—In anterolateral view, the petrosal is also teardrop-shaped, with its tip pointing dorsomedially. It is greatly convex anterolaterally. Its dorsal and lateral parts have a rough surface where they underlie the squamosal. Its anterior part is roughly triangular and smooth-surfaced, here designated as the anterior lamina. Within the suture between the posterior margin of the anterior lamina of the petrosal and the squamosal lies an intramural canal, that is, the ascending canal as described in the section entitled "Squamosal" earlier in this paper. The actual sutural line between the anterior lamina and the squamosal marks the anterior rim of the ascending canal. In many specimens in which the squamosal is broken off, the ascending canal is represented by a deep groove on the anterolateral surface of the petrosal. The canal connects both with the post-temporal canal dorsally and with the prootic canal foramen ventrally. The ascending canal runs anterodorsally to enter the cranial cavity through the foramen of the ascending canal and continues as the internal parietal groove on the cranial surface of the parietal (see also the description in the section entitled "Parietal"). Ventrally, the ascending canal opens at the prootic canal foramen, which faces ventrally and is anterior to the fenestra vestibuli. However, the specimen has to be rotated 90 degrees to the ventral view in order to see both the prootic canal foramen and the fenestra vestibuli.

Just where the ascending canal turns downward to open at the prootic canal foramen, there is a gap between the squamosal and the posterior process of the alisphenoid. Thus the gap exposes a very short segment of the ascending canal as an open groove. At the level at which the groove comes to be concealed laterally by the posterior process of the alisphenoid, the ascending canal becomes confluent with the prootic canal. Here the prootic canal emerges from a ventrally-facing foramen, the postglenoid foramen. The postglenoid foramen is situated immediately posterolateral to the foramen masticatorium, and at the dorsal edge of the ascending canal. Immediately anterodorsal to the postglenoid foramen is another

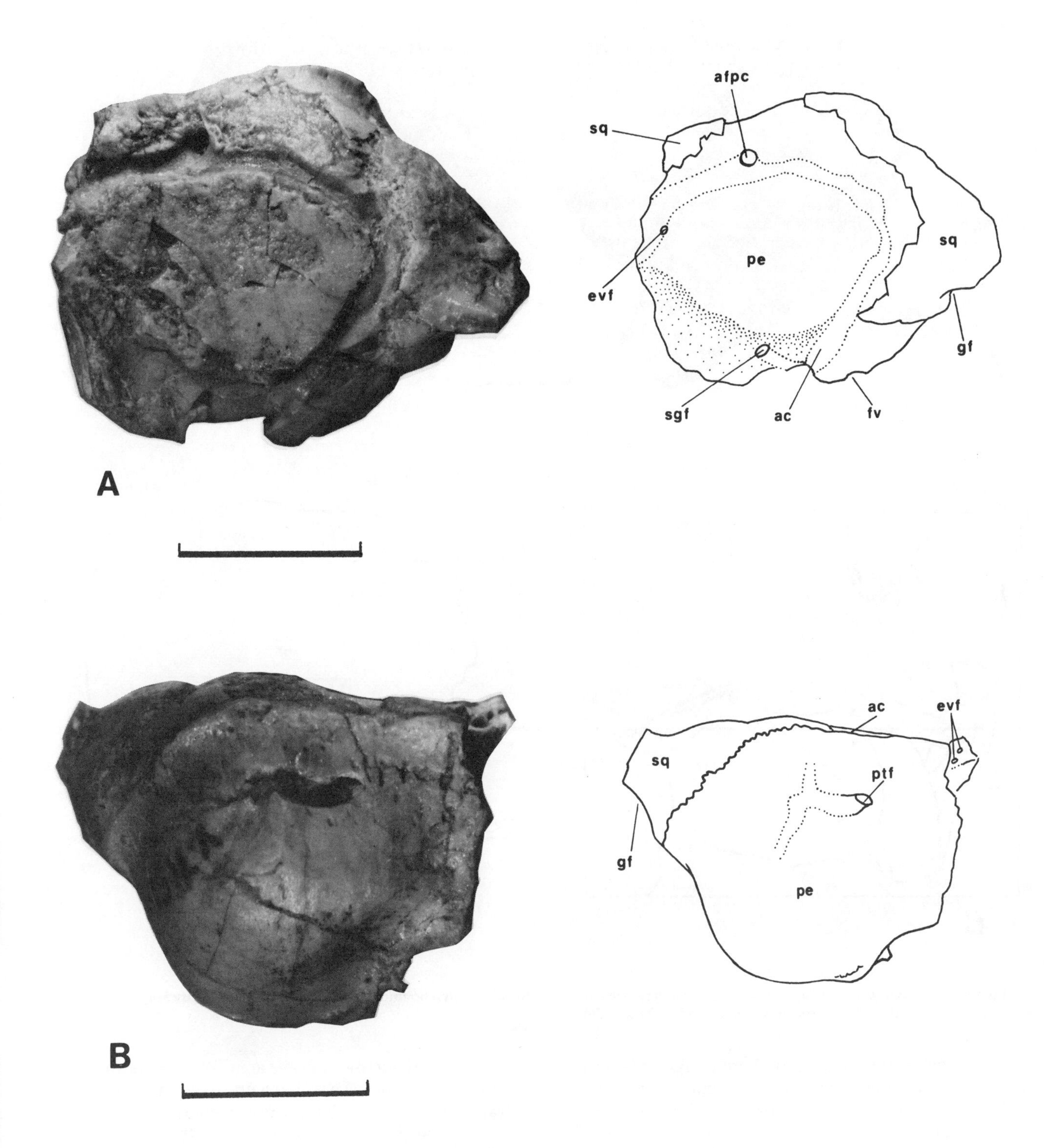

Figure 20. *Lambdopsalis bulla*, V7151.75, left petrosal. *A*, anterolateral view; and *B*, occipital (posterior) view.

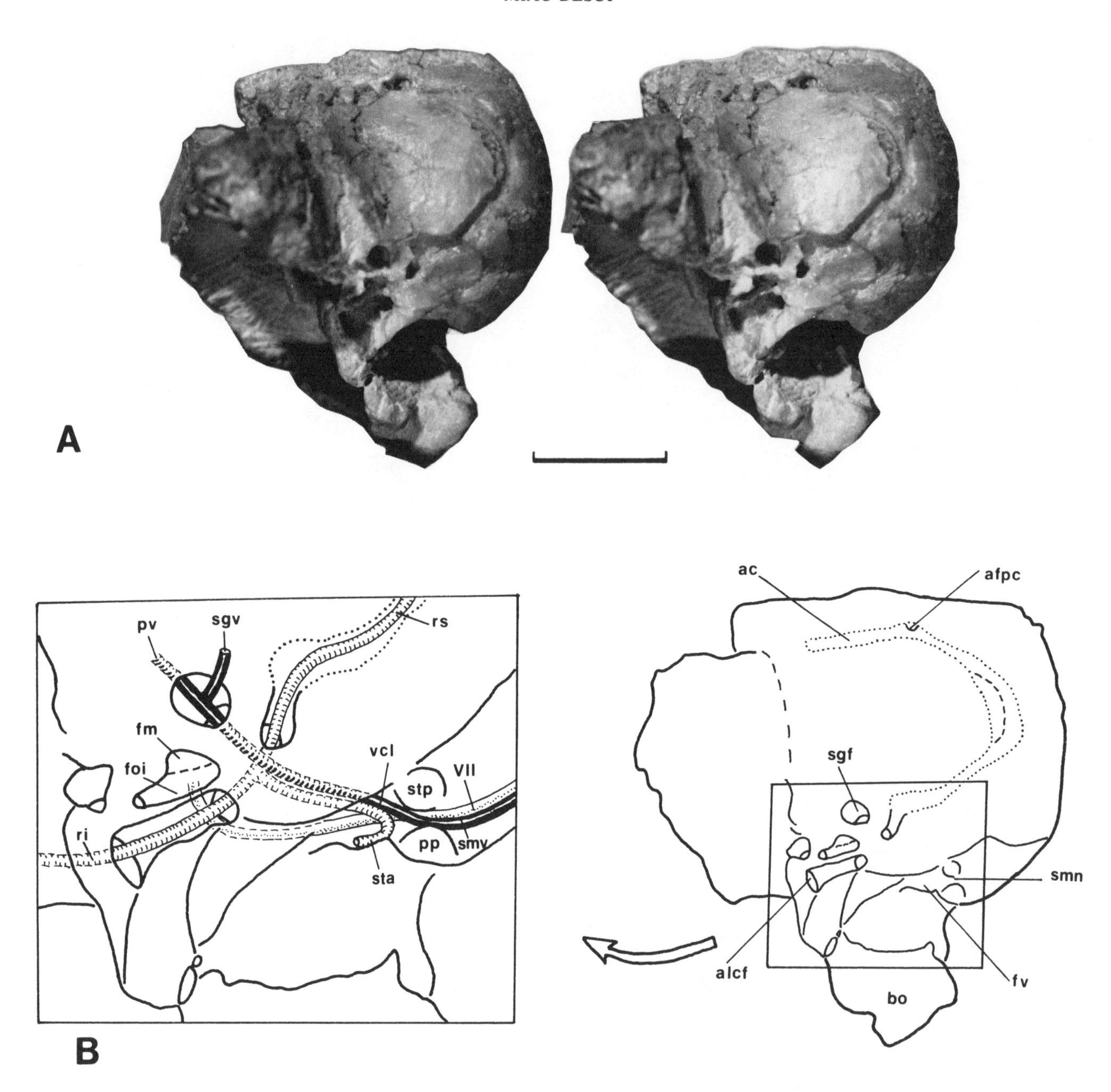

Figure 21. ***Lambdopsalis bulla*, V7151.57, anterolateral view of left petrosal. *A*, stereophotograph; and *B*, explanatory drawing with reconstruction of some unfossilized structures (see also Figs. 29 and 30).**

foramen, facing anteroventrally, here designated as the supraglenoid foramen. The canal connecting the supraglenoid foramen dorsally and the postglenoid foramen ventrally is part of the prootic canal. The supraglenoid foramen is dorsolateral to the foramen masticatorium. In specimens IVPP V7151.61 and V7151.94, the suture between the alisphenoid and the anterior lamina of the petrosal can be seen lying between the anterior rim of the supraglenoid foramen and the posterior rim of the foramen masticatorium. Posterior to the supraglenoid foramen, the anterior lamina of the petrosal rises as an anterodorsally-extended ridge, anterior to which is a roughly triangular and narrow concave area. Within the anterior lamina in this concave area lies a dorsoventrally-extended intramural canal, which is the dorsal continuation of the prootic canal. It can be traced through (by using a bristle) from the supraglenoid foramen dorsally to its cranial end near the anterodorsal corner of the subarcuate fossa. After emerging from the postglenoid foramen and becoming confluent with the ascending canal, the prootic canal continues ventrally to enter the facial sulcus. The foramen from which the prootic canal finally emerges is here designated as the tympanic foramen of the prootic canal, or simply the prootic canal foramen. This fora-

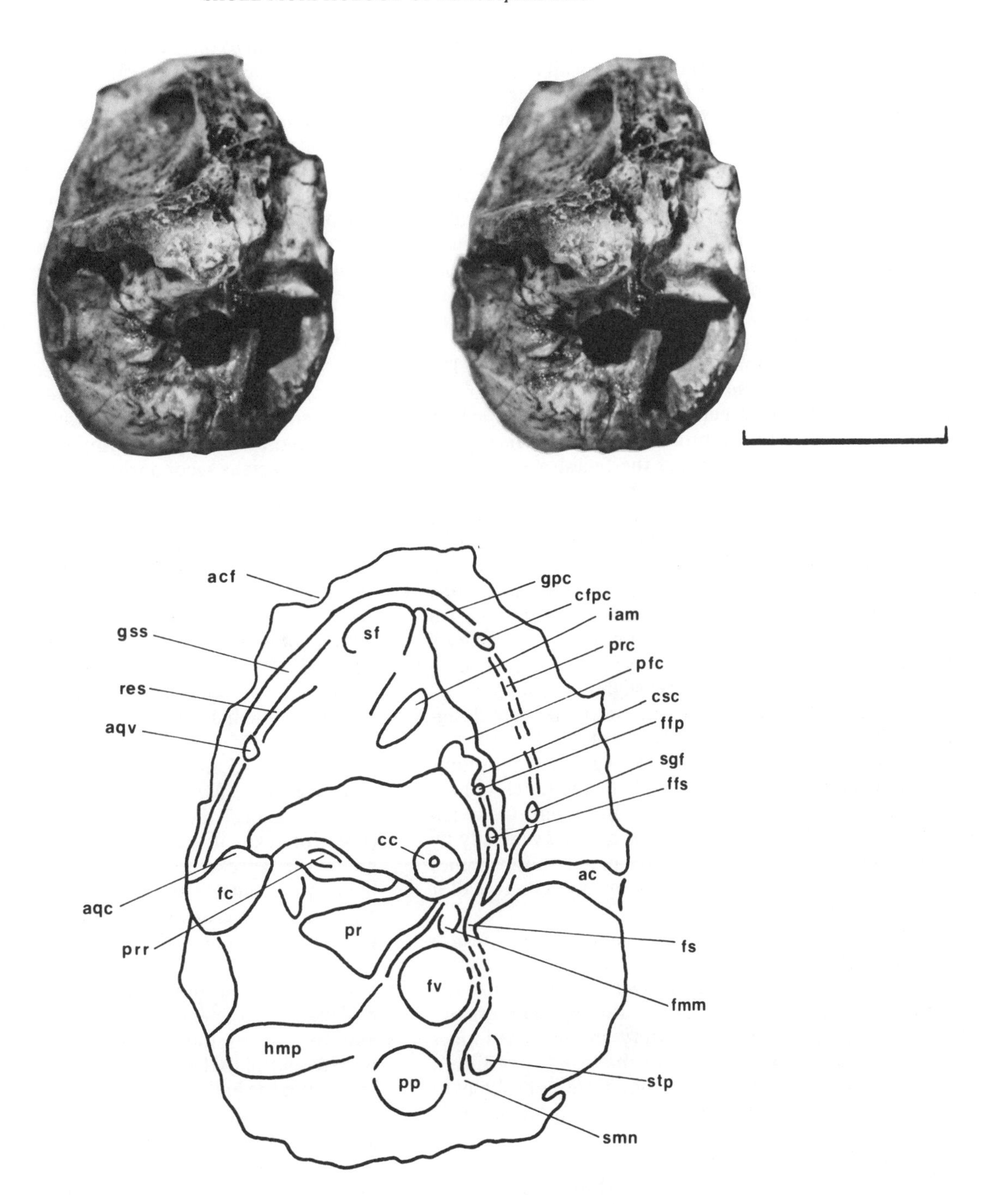

Figure 22. ***Lambdopsalis bulla*, V7151.58, ventral stereoscopic view of left petrosal.**

men is anteromedial to the fenestra vestibuli, and can be seen only in ventral view.

Also visible in lateral view, but often concealed by the posterior process of the alisphenoid, an anteroventrally-directed canal extends from the facial sulcus to the foramen immediately below the foramen ovale inferium. This canal is designated as the canal of the maxillary artery. The canal continues anteriad into the alisphenoid canal (see also the description and discussion of the alisphenoid canal in the section entitled "Pterygoid"). Anterodorsal to the anterior opening of the canal of the maxillary artery and anteroventral to the foramen ovale inferium is a small anteroventrally-facing foramen, here designated as the hiatus Fallopii, which opens into the vidian groove (see also the description and discussion of the pterygoid canal in the section entitled "Pterygoid").

Ventral aspect of petrosal.—In ventral view, the outline of the petrosal is roughly circular. The promontorium is moderately large and convex, but comparatively short. It is fairly straight, and directed anteromedially. The promontorium is in sutural contact with the basisphenoid anteromedially, with the posterior process of the alisphenoid laterally, and with the basioccipital medially. In several specimens (IVPP V7151.60, V7151.64, V7151.66, V7151.67, and V5429.8) in which the surface of the promontorium is broken, a straight cochlear endocast is well exposed. Posteriorly, there are two subequally-sized fenestrae flanking the sides of the promontorium. The larger one, at the posteromedial side of the promontorium, faces posteroventrally and is here designated the fenestra cochleae. At the anteromedial rim of the fenestra cochleae, a groove leads from the fenestra cochleae medially into the cranial cavity through a foramen. The groove is designated the aquaeductus cochleae, the foramen is the jugular foramen. The smaller fenestra, which lies posterolateral to the promontorium and faces ventromedially, is here designated the fenestra vestibuli.

On the medial side of the promontorium and anterior to the fenestra cochleae is a fossa-like concave area. This was described as the promontorium recess by Miao and Lillegraven (1986). Posterolateral to the fenestra vestibuli is a distinct rounded process, here designated the paroccipital process. Posterolateral to the fenestra cochleae and medial to the paroccipital process is an oval recess, designated the "hyoid muscle pit." Anterior to the paroccipital process is another slightly smaller process, lying anterolateral to the fenestra vestibuli, here designated the styloid process. There is a notch between the styloid and paroccipital processes, which is continuous anteromedially with the facial sulcus; the notch is here designated the stylomastoid notch. In at least one specimen (IVPP V7151.57), the notch is bridged ventrally by a tiny bony outgrowth of the petrosal, and thus becomes a foramen. In this case, it is more appropriate to call it the stylomastoid foramen.

Two grooves are seen in the vicinity of the fenestra vestibuli. The shorter one extends anteromedially from the posteromedial rim of the fenestra vestibuli, and lies along the lateral side of the promontorium. This groove is designated as the stapedial artery groove. The longer groove flanks the anterior rim of the fenestra vestibuli, extending from the stylomastoid notch anteromedially to a foramen just posteroventral to the foramen ovale inferium. The groove is here designated as the facial sulcus, and the foramen as the foramen facialis secondarium. The anteromedial segment of the facial sulcus, as well as the foramen facialis secondarium, is concealed by the posterior process of the alisphenoid in straight ventral view. The specimen has to be tilted forward about 30 degrees in order to see these structures. However, in the holotype (IVPP V5429), due to the partly broken posterior process of the alisphenoid, both the anterior part of the facial sulcus and the foramen facialis secondarium are exposed. A distinct fossa is present between the stapedial artery groove and the facial sulcus, just medial to the fenestra vestibuli and posteroventral to the foramen facialis secondarium, here termed the fossa muscularis major. There is not a well-defined fossa on the posterior side of the fenestra vestibuli and immediately anteromedial to the paroccipital process. Anterior to the fenestra vestibuli, the transversely-extended posterior process of the alisphenoid forms a distinct concave area on its ventral surface. The concave area is designated as the epitympanic recess, which is bounded both by an anterior ridge and a posterior ridge. Medial to the epitympanic recess and at the junction of the promontorium and the posterior process of the alisphenoid, there is a short, narrow, and anteromedially-directed groove. The groove is roughly parallel and anterolateral to the stapedial artery groove.

In the type specimen (IVPP V5429), in which the posterior process of the alisphenoid is partly broken, a tiny bony bridge can be observed lying between the posteroventral edge of the foramen ovale inferium and the anterior rim of the foramen facialis secondarium, here designated as the tegmen tympani. In other words, the foramen facialis secondarium is within the tegmen tympani, and connects the facial sulcus with the posterior part of the cavum epiptericum.

Anteromedial aspect of petrosal.—In anteromedial or cranial view, the outline of the petrosal is roughly pear-shaped. The cranial surface of the petrosal is overall concave, with two distinct fossae. The dorsal one, designated as the subarcuate fossa, is in the mastoid part of the petrosal. It is roughly rounded, shallow, smooth-surfaced, and bounded by ridges. The anterior ridge marks the medial wall of the prootic canal. The dorsal and posterior ridges separate the subarcuate fossa from a groove dorsal and posterior to the fossa. This groove is convex dorsally, with its shorter anterior segment descending into the prootic canal and with its longer posterior segment descending along the posterior margin of the petrosal to emerge through the jugular foramen to the outside of the cranial cavity. The anterior segment of the groove is designated as the groove of the prootic sinus, and the posterior segment of the groove as the groove of the sigmoid sinus. At the apex of the dorsally convex groove, there are several tiny foramina, designated as the emissary venous foramina. Along the anterior rim of the groove of the sigmoid sinus and posteroventral to the subarcuate fossa is a foramen which opens into the hollow space inside the petrosal, here identified as the aquaeductus vestibuli. A short but distinct groove extends from the aquaeductus vestibuli anterodorsally to end at a recess posterodorsal to the subarcuate fossa. The recess is here designated the recess of the endolymphatic sac, which is immediately ventral to, and partly overlaps, the groove of the sigmoid sinus.

Ventral to the subarcuate fossa is a much smaller and deeper fossa, here designated the internal auditory meatus. It is in the petrous part of the petrosal. There are three separate foramina within the internal auditory meatus, divided by a transverse crest.

Dorsal to the transverse crest, there are two foramina. The larger one is dorsally placed and opens into the hollow space within the petrosal, here designated as the foramen for the vestibular nerve (VIII). The smaller foramen, here designated as the cranial foramen of the aquaeductus Fallopii for the facial nerve (VII), is

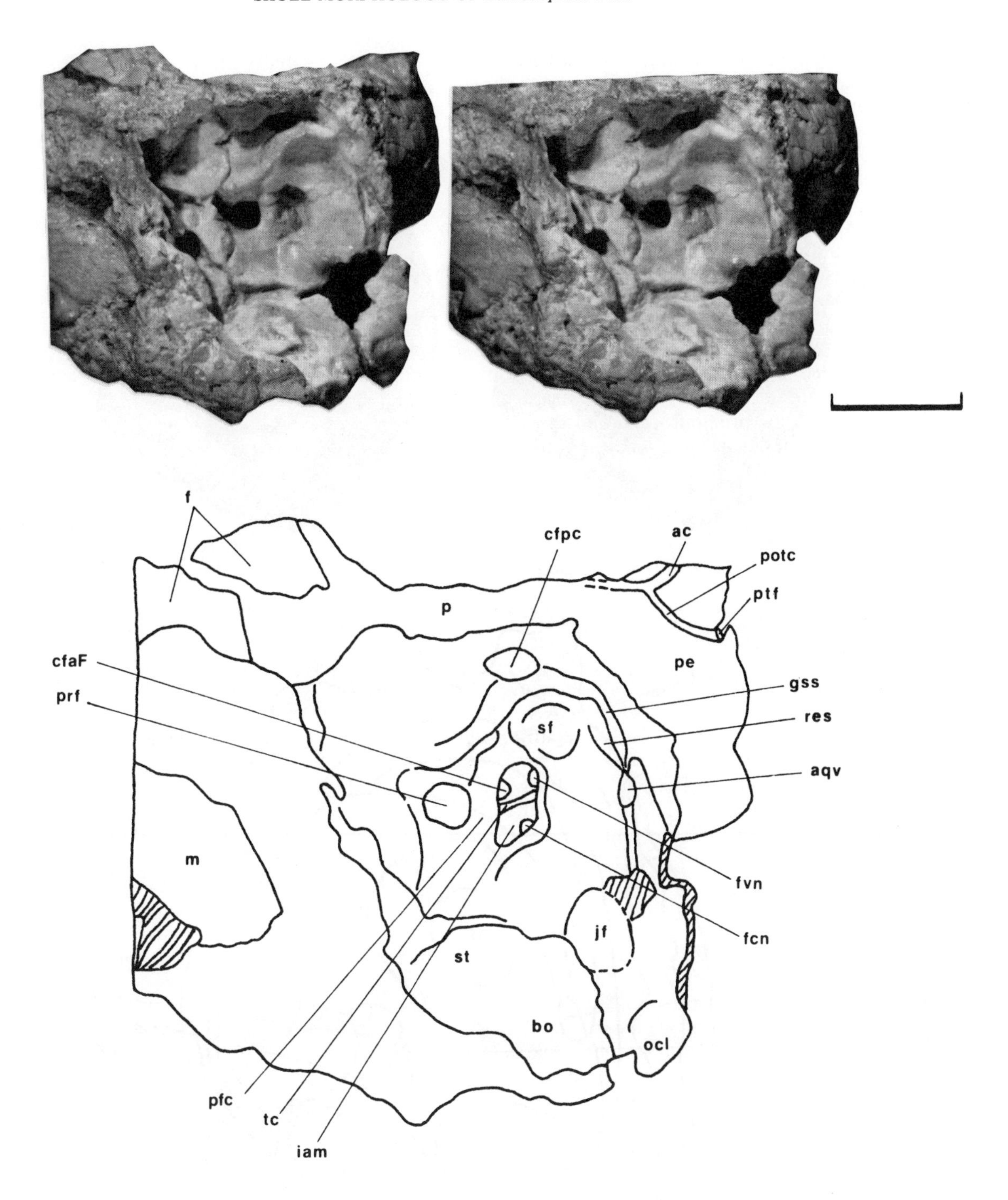

Figure 23. ***Lambdopsalis bulla*****, V7151.56, stereophotograph, anteromedial (cranial) view of right petrosal.**

anteroventral to the foramen for the vestibular nerve, and can be traced through its tympanic opening into the posterior part of the cavum epiptericum. The tympanic opening of the aquaeductus Fallopii facing toward the cavum epiptericum is designated the foramen facialis primitivum. The bony canal between these two foramina is designated the aquaeductus Fallopii (or facial canal). The bone enclosing the foramen facialis primitivum and forming the posterior margin of the prootic foramen (as described earlier in the paper) is designated the prefacial (or suprafacial) commissure. If one looks posterolaterally through the prootic foramen into the posterior part of the cavum epiptericum, a small fossa is seen to lie between the prefacial commissure medially and the tegmen tympani laterally; it is here designated the cavum supracochleare. Anterior to the cavum supracochleare is a much larger fossa, here termed the semilunar fossa. In the tegmen tympani, a very short canal connects the

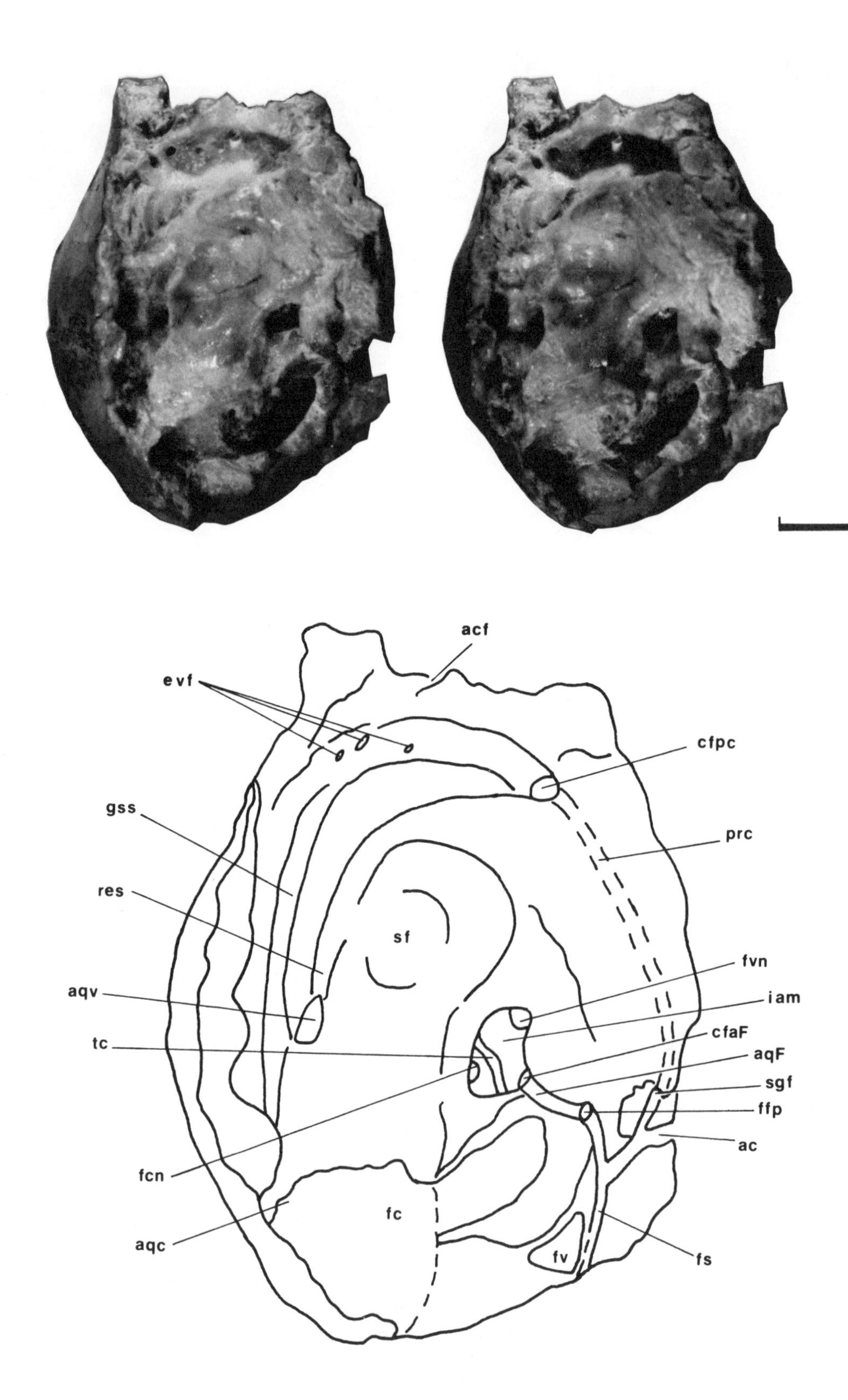

Figure 24. ***Lambdopsalis bulla*, V7151.75, anteromedial stereoscopic view of left petrosal.**

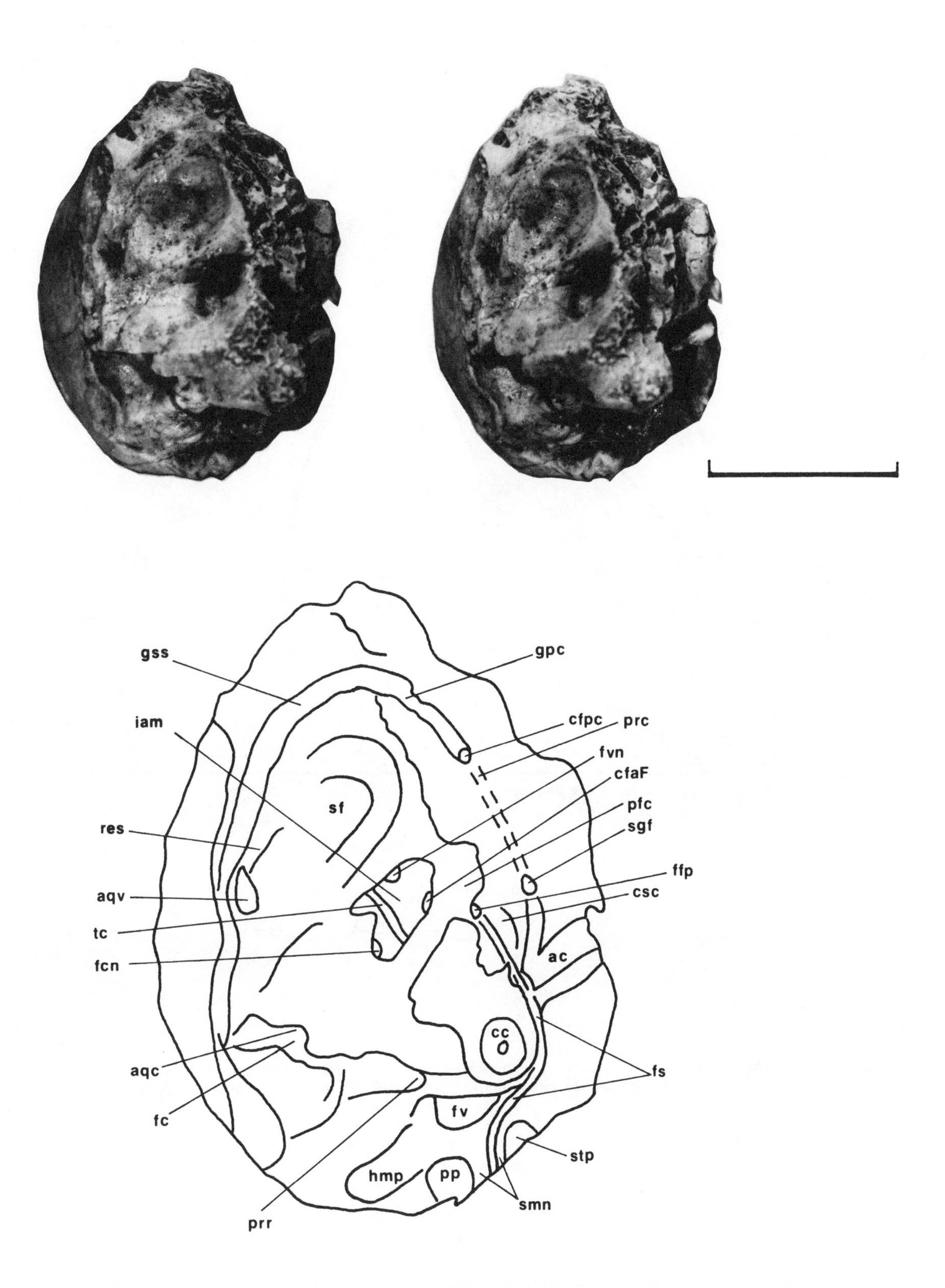

Figure 25. ***Lambdopsalis bulla*, V7151.58, stereophotograph, anteromedial view of left petrosal.**

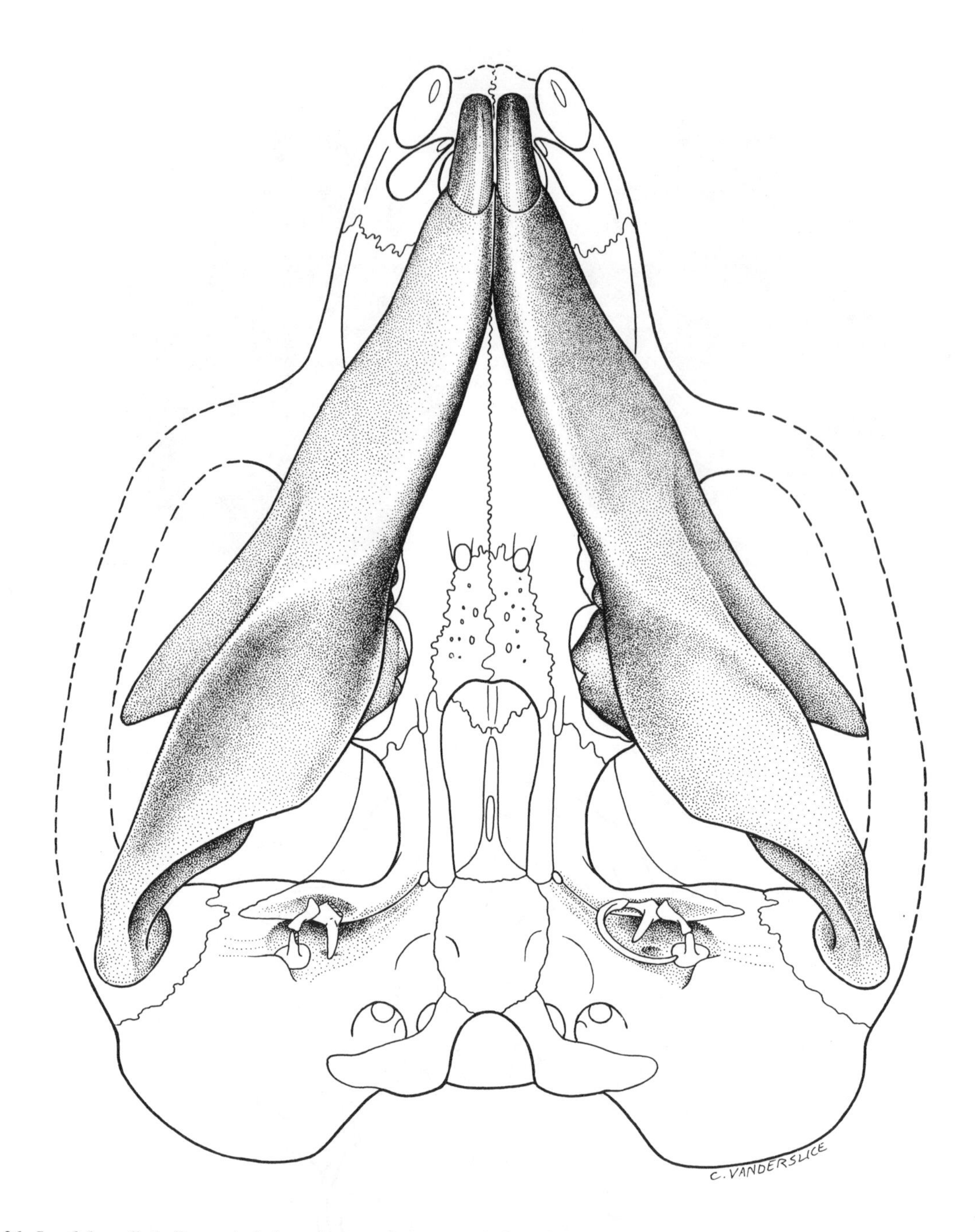

Figure 26. ***Lambdopsalis bulla*****, ventral view of composite reconstruction of skull with articulated mandibles. Ear ossicles on right side were drawn from actual specimen (V7151.80, see also Miao and Lillegraven, 1986); ear ossicles and tympanic bone on left side are conjectural.**

cavum supracochleare medially with the facial sulcus ventrolaterally. As described above, the tympanic opening of the short canal is the foramen facialis secondarium.

Ventral to the transverse crest of the internal auditory meatus is another small, posterolaterally-placed foramen, here designated the foramen for the cochlear nerve (VIII). This foramen opens into the internal chamber of the promontorium. Ventral to the prootic foramen and anteroventral to the internal auditory meatus is the smooth, medial (or cranial) surface of the promontorium. At the posterior margin of the medial surface of the promontorium, and posteroventral to the foramen for the cochlear nerve, is a distinct notch, which can be traced laterally toward the fenestra cochleae. This notch is the cranial opening of the aquaeductus cochleae (as described earlier in this section).

Region of middle ear.—The region of the middle ear, and especially the three ear ossicles, of *Lambdopsalis* was described elsewhere by Miao and Lillegraven (1986). A few supplementary notes about the middle ear region have been added here in the description of the ventral aspect of the petrosal. A tympanic bone has not yet been found.

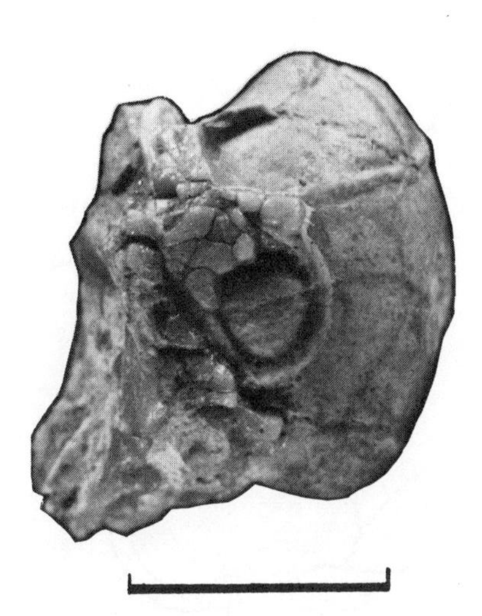

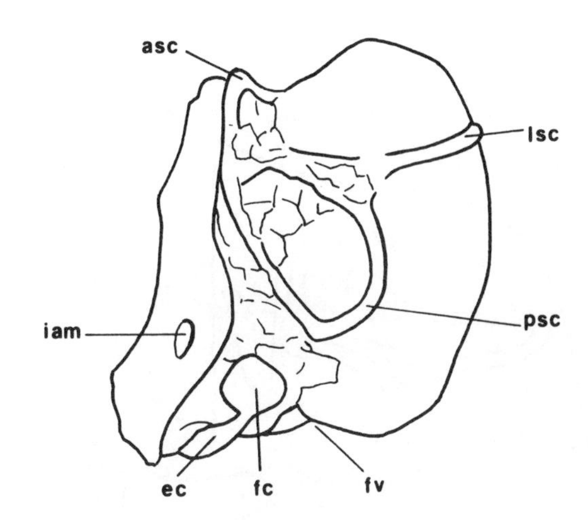

Figure 27. ***Lambdopsalis bulla*, V7151.60, posterior view of right petrosal. With parts of bone removed, showing endocasts of cochlea, three semicircular canals, and vestibule.**

Structures of inner ear.—Structures of the inner ear are seen best in endocasts of the osseous labyrinth plus its surrounding space within the petrosal. The endocast completely fills the spheroidal, hollow space inside the petrosal. The surface of the endocast is mostly smooth, except where the canals, bulges, and fenestrae exist. The endocast looks roughly like a strawberry in an upside down position with its "stalk" anteroventrally-directed. The "stalk" represents the endocast of a straight cochlea. The "strawberry" itself is the endocast of the vestibular apparatus plus the space surrounding it. It is impossible to delimit the utricle and the saccule, and thus they are here described together as the vestibule. However, the three semicircular canals are clearly observable. Each of the semicircular canals has an expanded part at one end, just before the canals enter the vestibule. The expanded part is designated the ampulla. The anterior semicircular canal is intermediate in length among the three, and lies in a vertical plane. It is gently curved with its convexity upward, and much less than a half-circle in arc. The anterior semicircular canal lies in an arcuate eminence dorsal to the grooves of both the prootic and the sigmoid sinuses, not in the dorsal ridge of the subarcuate fossa. The ampulla of the anterior semicircular canal lies at the anterior end (at a level of the cranial opening of the prootic canal), and immediately dorsal to the ampulla of the lateral semicircular canal. The most posterior part of the anterior semicircular canal shares a common stem, termed the crus commune, with the dorsal end of the posterior semicircular canal. They open into the vestibule by one common aperture at a location immediately dorsal to the aquaeductus vestibuli. The posterior semicircular canal is the shortest, and also lies in a vertical plane perpendicular to the long axis of the skull. Its convexity points posterolaterally. Its ampulla is on the ventromedial extremity, and opens into the vestibule just posterolateral to the fenestra cochleae. The lateral semicircular canal is the longest, and lies in a roughly horizontal plane but with its anterolateral convexity slightly tilted upward. Its ampulla is at the anterior end, and immediately ventral to the ampulla of the anterior semicircular canal. The posterior end of the lateral canal opens into the vestibule ventrolateral to the crus commune. Both the posterior and the lateral semicircular canals are more than a half-circle in arc. The vestibule, being the site of the fenestra vestibuli, is disproportionately enlarged.

Discussion

The mammalian petrosal is a cartilage replacement bone. The petrosal, especially its petrous part, is generally more dense than other cranial bones, and thus stands a better chance of being preserved as a fossil. Despite Kermack and others' (1981, p. 80) statement that "It would be reasonable to expect that such a bone would never survive the mechanical stresses of fossilization", they themselves found that "it is unexpected—indeed most fortunate—" that many specimens of petrosal of *Morganucodon* are almost completely preserved. In fact, quite contrary to Kermack and others' unusual expectation, the petrosal is usually better represented, better preserved, and thus better known than any other element among the cranial materials of early mammals (*e. g.*, Simpson, 1928, 1937; Kermack and Mussett, 1958; Kermack, 1963; Hahn, 1969, 1978; Kermack and others, 1981; Crompton and Sun, 1985; and Kielan-Jaworowska and others, 1986).

Thanks to Simpson's (1937) and Kielan-Jaworowska and others' (1986) excellent descriptions, the petrosal of multituberculates is now adequately understood. The petrosal of *Lambdopsalis* accords well with the general structural pattern as described by these earlier authors, but there are detailed differences in relative extent and position of certain structures. To avoid repetition, the differences rather than the similarities in structural details and/or interpretations will be stressed in the following discussion.

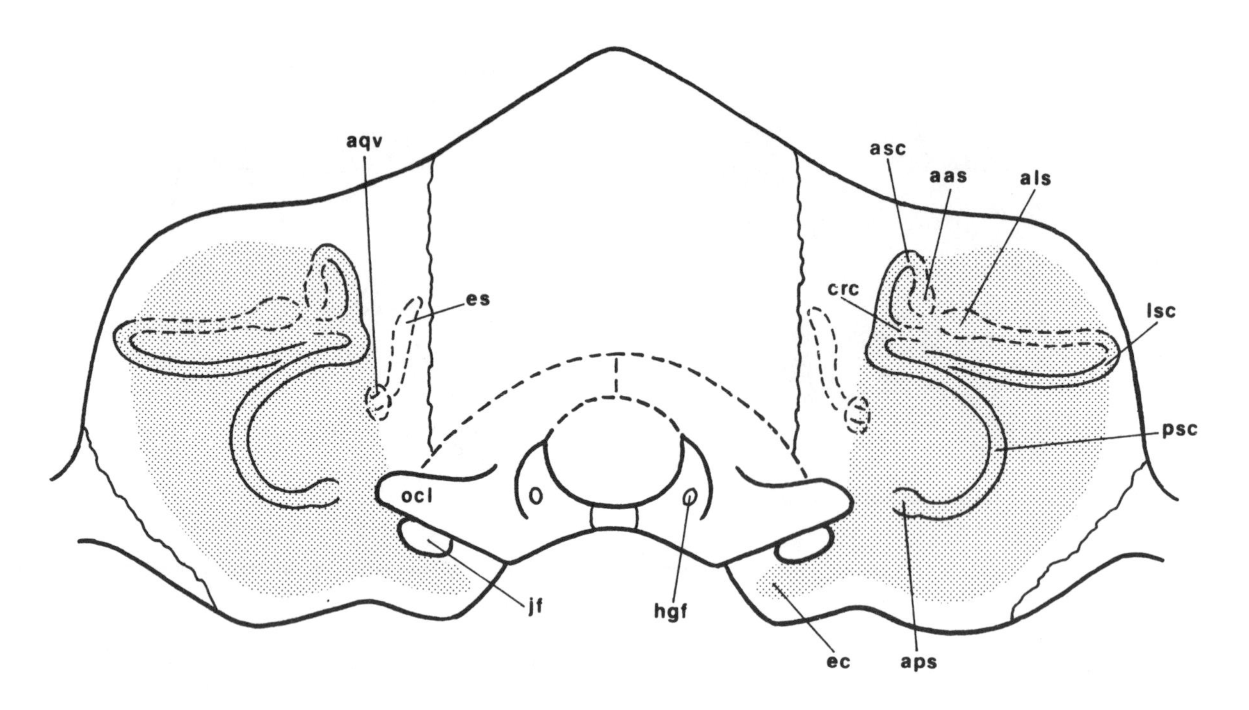

Figure 28. ***Lambdopsalis bulla*, occipital view of composite reconstruction of skull. Shaded area shows endocast of expanded vestibular apparatus.**

Anteromedial inclination of anterior lamina.— Kielan-Jaworowska and others (1986) noted that two types of braincase structure may be recognized in comparison of the Mongolian taeniolabidoid skulls with isolated petrosals from the Hell Creek Formation of Montana. That is, the anterior lamina of the petrosal is more medially inclined in various degrees in all available cranial materials of taeniolabidoids. They also noted that the larger the size of the animal, the more the anteromedial inclination of the anterior lamina. Kielan-Jaworowska and others (1986, p. 555) concluded: "... that the anteromedial inclination of the anterior lamina of the petrosal may be related to a specialization of the jaw muscles in these larger forms, in which the fourth premolars tend to be reduced and the main cutting action, probably gnawing, was done by the very enlarged, self-sharpening, incisors, a similar pattern to that of rodents." Their functional explanation, though appearing plausible, needs to be corroborated in light of discoveries in *Lambdopsalis*.

As described above, the anteromedial inclination of the anterior lamina is carried so far that much of the "anterior lamina" (as defined by Kielan-Jaworowska and others, 1986) in *Lambdopsalis* faces anteriorly. In other words, much of the "anterior lamina" in *Lambdopsalis* is parallel to the transverse plane of the skull. *Lambdopsalis* is large among the known taeniolabidoids, but it is not the largest. Thus the great anteromedial inclination of the anterior lamina seen in *Lambdopsalis* cannot be explained satisfactorily by Kielan-Jaworowska and others' (1986) interpretation. Rather, the answer may lie in the great expansion of the vestibular apparatus. The extraordinary expansion of the bony vestibular apparatus may have "pushed" (so to speak) the anterior lamina anteromedially, and "squeezed" the anterior lamina out of its usual place. In addition to the anteromedial inclination of the anterior lamina, this great expansion of the vestibular apparatus also may have caused a number of other unusual changes in the following structures at its proximity as seen in *Lambdopsalis*: (1) the fenestra vestibuli faces anteromedially, rather than normally facing laterally or posterolaterally; (2) the epitympanic recess is transversely oriented, rather than normally anteromedially/posterolaterally oriented (*e. g.*, in *Ptilodus*); (3) much of the squamosal faces anteriorly or lateroanteriorly, rather than normally facing laterally; (4) the subarcuate fossa becomes very shallow; and (5) both the postorbital temporal region and the postpalatal basicranium are greatly shortened.

The interpretation offered herein is strengthened further by discovery of a pea-sized hollow space inside the petrosal of ?*Catopsalis joyneri* from the Bug Creek Anthills; this is the same taxon (and is from the same locality) as the one that Kielan-Jaworowska and others (1986) described (Luo Zhexi, *personal communication*; and my personal observation). Similar to the hollow space inside the petrosal of *Lambdopsalis*, this one presumably also housed the vestibular apparatus. It shows similar expansion, but to a lesser degree. Plate 1 of Kielan-Jaworowska and others (1986) clearly shows that the petrosal of ?*Catopsalis joyneri* has a strongly convex occipital surface and a great medial inclination of the anterior lamina, presumably due to the expansion of the bony vestibular apparatus inside the petrosal. Therefore, I predict that, in taeniolabidoids, the greater the expansion of the vestibular apparatus, the greater the medial inclination of the anterior lamina.

Problem of anterior lamina of petrosal.—So far, my usage of the term "the anterior lamina of the petrosal" has followed closely that of Kielan-Jaworowska and others (1986). It was defined in multituberculates as "a large component of the lateral aspect of the braincase", parallel to an "extension of petrosal forward as lateral wall of cavum epiptericum to form part of wall of temporal fossa" in advanced therapsids and nontherian mammals (Kielan-Jaworowska and others, 1986, p. 578). In practice, the lateral aspect of the anterior lamina is recognized as "the large, roughly triangular surface of the anterior lamina of the petrosal, bounded posteriorly and dorsally by the ascending canal" (*ibid.*, p. 530). However, the underlying problems with the definition (as well as the practice) are the facts that: (1) much of so-called "anterior lamina of the petrosal" is the lateral aspect of the petrosal itself, but not a forward extension of the bone; and (2) the "lamina" is *posterior* to rather than forming the lateral walls of both the cavum epiptericum and the temporal fossa. Only a very narrow bony strip anterior to the prootic canal can be claimed as the forward extension of the petrosal. Yet this would exclude the large, roughly triangular surface between the prootic and the ascending canals from being part of the anterior lamina. I stress again that the enlargement of this roughly triangular surface seems to encrust the expanded vestibular apparatus. Therefore, descriptively, it is more desirable to call only the bony strip between the alisphenoid/petrosal suture and the prootic canal the anterior lamina. However, for the sake of effective communication, I have followed Kielan-Jaworowska and others (1986) closely in the recognizing and describing the anterior lamina, but I do not consider this practice in accordance with the proper definition of the anterior lamina of the petrosal as originally coined by Watson (1916) for monotremes.

It also should be stressed that, for the developmental reasons argued by Presley (1981), Kielan-Jaworowska and others (1986) did not deem that a large contribution from the anterior lamina to the lateral wall of the braincase has any fundamental phylogenetic significance. In addition, if, as I have argued thus far, much of the so-called "anterior lamina" only represents part of the lateral aspect of an expanded petrosal and does not contribute significantly to the formations of either the lateral wall of the cavum epiptericum or that of the temporal fossa, then the notion of multituberculates possessing a nontherian, monotreme-like braincase pattern itself is probably a myth.

Post-temporal canal, ascending canal, and orbito-temporal vascular system.—Kielan-Jaworowska and others (1986) made an important distinction in structural details of the petrosal between *Catopsalis,* with a small post-temporal fossa, and genera with a large post-temporal fossa such as *Nemegtbaatar, Chulsanbaatar, Kamptobaatar,* and *Sloanbaatar*. As described earlier, the post-temporal fossa in *Lambdopsalis* is even smaller proportionately, and more dorsally-placed, than that in *Catopsalis*. Also similar to *Catopsalis*, but different from the genera with a large post-temporal fossa, *Lambdopsalis* has no major fenestration (except a few tiny emissary venous foramina) in its subarcuate fossa. Added to this listing is that the post-temporal canal vessels (*e. g.*, arteria diploetica magna) in *Lambdopsalis* were intramural rather than endocranial, just as in *Catopsalis*. The post-temporal canal vessels (*e. g.*, arteria diploetica magna) probably sent only a tiny branch communicating with the subarcuate fossa. The main vessels entered the ascending canal to join the ascending canal vessels (*e. g.*, ramus superior of stapedial artery). Then the ascending canal vessels both ascended into the internal parietal groove and descended to connect the stylomastoid and/or the postglenoid vessels. This is what Kielan-Jaworowska and others (1986) called the orbito-temporal vascular system. The petrosal of *Lambdopsalis* shows firm osteological evidences in favor of their reconstruction. Particularly, *Lambdopsalis* accords well with their restoration of the orbito-temporal vascular system of *Catopsalis*. Incidentally, due to the extremely dorsal position of the post-temporal canal in *Lambdopsalis*, much of the course of the ascending canal is ventral to the post-temporal canal and, therefore, should be more properly called the descending canal. However, for the consistency of the terminology, the term "ascending canal" is adopted in this paper.

Similar bony markings (be it a canal or groove) have been reported also in various cynodonts and nontherian mammals, and interpreted as being entirely or chiefly venous (*e. g.*, Watson, 1911; Kemp, 1979; and Kermack and others, 1981). As Kielan-Jaworowska and others (1986, p. 572) clearly pointed out: "In modern mammals both veins and arteries can be found in portions of this course", that is, the orbito-temporal vascular system. However, these authors (1984, 1986) suggested that the main occupants of these canals (or sometimes grooves) were arteries rather than veins. They listed at least three reasons in developing their argument, namely: (1) striking similarity to the arteria diploetica magna in *Tachyglossus*; (2) arteries, not veins, related consistently to the anterior part of the system in many eutherians (contrary to W. K. Parker, 1885); and (3) in modern mammals, such arteries and their venae comitantes passing beside, rather than joining directly, the bony markings for the dural sinuses.

This view receives strong support from Wible's (*in press*) study on the cranial vascularization in modern sauropsids. It was on the cranial venous pattern of *Sphenodon* (see O'Donoghue, 1920) and modern lizards (Bruner, 1907; and Shindo, 1915) that the initial reconstruction of the venous pattern in cynodonts and nontherian fossil mammals was based (Watson, 1911, 1920). However, Wible suggested that actually the two are not directly comparable. He further pointed out that in modern sauropsids it is the stapedial artery and its branches that lie directly on the braincase, whereas the major extracranial veins (except the vena capitis lateralis) are laid more superficially within the adductor muscles. In other words, those extracranial veins are too superficially-placed to be associated with grooves on the cranial bones.

The bony markings on the surfaces of the petrosal (as well as the parietal) in *Lambdopsalis* are almost

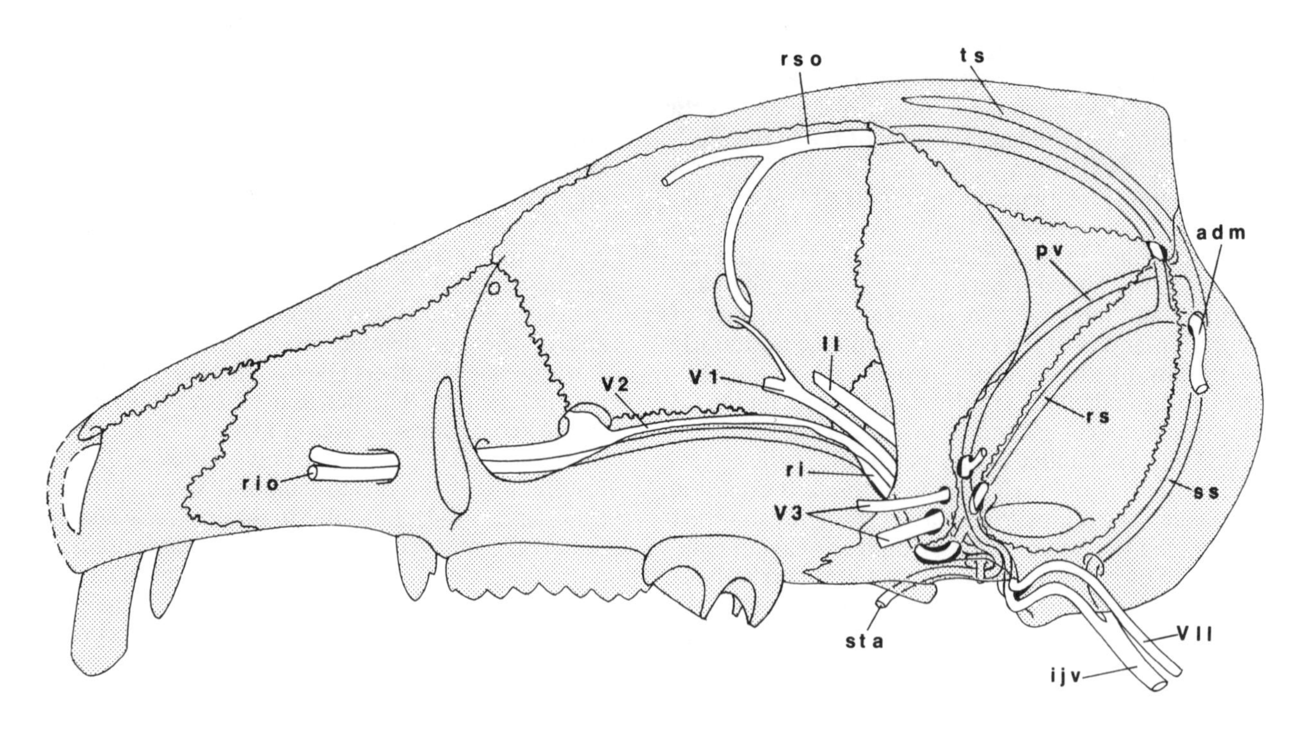

Figure 29. Reconstruction of orbito-temporal vascular system, dural sinus system, and certain cranial nerves in *Lambdopsalis bulla*.

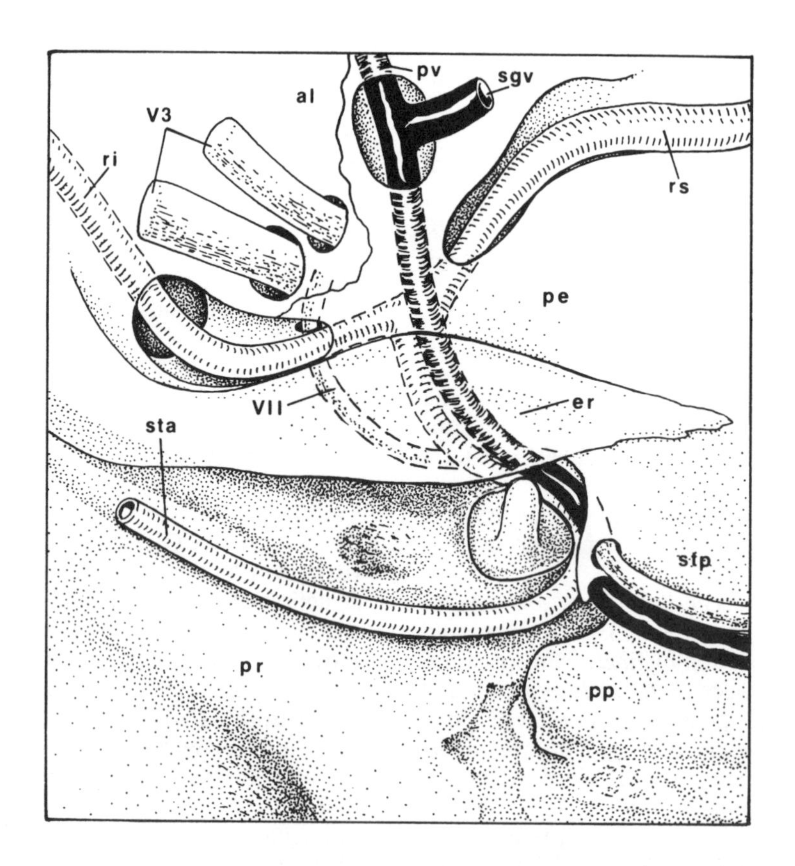

Figure 30. Oblique (ventrolateral) view of ear region of *Lambdopsalis bulla*, showing reconstruction of some major vascular and nervous structures as deciphered from osteologic evidence.

identical to those of *Catopsalis* as described by Kielan-Jaworowska and others (1986). For that reason, it seems reasonable to apply reconstructions of the orbitotemporal vascular pattern of *Catopsalis* to *Lambdopsalis*. However, Kielan-Jaworowska and others (1986) suggested that, although the anterior part of the orbitotemporal system did not differ between *Catopsalis* and *Nemegtbaatar* or *Chulsanbaatar*, ". . . one should perhaps think of occipital vessels in the latter two, and vessels in the stylomastoid or postglenoid regions in *Catopsalis*, as being the principal caudal connection" (p. 567). In *Lambdopsalis*, the ventral connection of the ascending canal vessels with the vessels in the postglenoid and/or stylomastoid region seems probable (as judged from the osteological evidence). However, the posterior connection of the post-temporal vessels with the occipital vessels also seems possible in *Lambdopsalis*. As described above, in occipital view of the petrosal in *Lambdopsalis* a distinct groove extends posterolaterally from the post-temporal fossa. Possibly, it indicates the presence of an arterial connection with the occipital artery, similar to that in *Ornithorhynchus* (see also Wible, 1987, Fig. 2A). However, Kielan-Jaworowska and others (1986, p. 568-569) pointed out: "As demonstrated by Bugge (1972, 1974), the arterial system is extremely variable in mammals, especially with respect to anastomoses between principal arteries (*i.e.* common, internal and external carotids) and distributaries supplying particular regions." Therefore, it cannot be overemphasized that alternative possibilities in restorations of vascular systems in fossils should always be left open.

Dural sinus system.—The venous sinus system has been reconstructed in many cynodonts (*e. g.*, Watson, 1911, 1920; Kühne, 1956; Cox, 1959; and Hopson, 1964), and in morganucodontids (Kermack and others, 1981). Although minor details of interpretation differ among these authors, a relatively constant pattern of the system has been recognized. This basic pattern of the venous sinuses for cynodonts and morganucodontids is well illustrated in Figure 1A and 1B of Wible (*in press*). For *Morganucodon*, see also Figure 76 in Kermack and others (1981). However, as noted immediately above, some of those vessels appear not to have occupied the bony markings as claimed by earlier authors. This implies that the previous restorations of the venous sinus system in cynodonts may be seriously doubted. It also should be noted that Kermack and others (1981) restored a dural sinus system for *Morganucodon* similar to that of *Ornithorhynchus*. Kermack and others (1981) also adopted terminology applied to various (some probably non-homologous) vessels in sauropsids. Unfortunately, "This has caused considerable misunderstanding in mammal-reptile comparisons" (Kielan-Jaworowska and others, 1986, p. 588). In the following discussion, terminology follows Butler (1967), and Kielan-Jaworowska and others (1986).

As reconstructed by Kielan-Jaworowska and others (1986), the dural sinus system in multituberculates had typical mammalian characteristics, and included the sagittal, transverse, prootic, and sigmoid sinuses, arising from the primary head vein (see also Butler, 1967). Judged from osteological evidence, *Lambdopsalis* had a similar system. The sagittal sinus occupied the median groove on the ventral surface of the parietal (see also the section entitled "Parietal"). Posteriorly, the sagittal sinus bifurcated into a transverse sinus on each side over the petrosals. The transverse sinus, in turn, bifurcated into a prootic sinus anteroventrally and a sigmoid sinus posteroventrally. The prootic sinus was contained by the groove of prootic sinus, and then passed downward into an essentially vertical, smooth-walled canal, the prootic canal. After entering the prootic canal, the prootic sinus probably became the prootic vein. The prootic vein met the supraglenoid vein near the supraglenoid foramen. The main trunk of the prootic vein continued downward to emerge through the postglenoid foramen and the prootic canal foramen to join the primary head vein in the facial sulcus. This junction also marks the point where the primary head vein was subdivided arbitrarily into a lateral segment posteriorly, called the lateral head vein (vena capitis lateralis), and a post-trigeminal segment anterodorsally, called the post-trigeminal vein. The foramen facialis secondarium marks the arbitrary boundary between the post-trigeminal vein and the medial head vein. Medial to the foramen and inside the semilunar fossa, the vein, though continuous with the post-trigeminal vein, became known as the medial head vein. The medial head vein, medial to the trigeminal ganglion, drained posteroventrally into the lateral head vein. The lateral head vein turned posterolaterally, and ran in the facial sulcus along with the facial nerve. The vein and nerve left the tympanic cavity through the stylomastoid notch. The lateral head vein, now known as the stylomastoid vein, drained into the internal jugular vein.

Interpretation of the venous sinuses in *Lambdopsalis* thus far is similar to that of the Mongolian multituberculates and ?*Catopsalis joyneri* as restored by Kielan-Jaworowska and others (1986). However, foramina for the supraglenoid, postglenoid, and prootic canal in *Lambdopsalis* are essentially aligned, in contrast to the triangular arrangement in genera described by Kielan-Jaworowska and others (1986). Therefore, there seems to have been no complicated venous tributary connections among the three foramina in *Lambdopsalis*. Nor is there any sign of presence of a tentorial sinus in *Lambdopsalis*, in contrast to the taxa studied by Kielan-Jaworowska and others (1986). Moreover, what these authors called the post-trigeminal canal is here equated in *Lambdopsalis* with the foramen facialis secondarium. I suggest that the main trunk (rather than the superficial petrosal branch) of the facial nerve left the cavum supracochleare through the foramen facialis secondarium. This problem will be discussed further in relation to the route of the facial nerve.

It should be noted that: (1) the prootic sinus plus the prootic vein in *Lambdopsalis* are equivalent to the vena cerebralis media as identified in *Morganucodon* (see Kermack and others, 1981); (2) the vena capitis lateralis of *Morganucodon* may represent more than the mere lateral segment of the primary head vein in multituberculates; and (3) in early mammals the prootic canal is invariably in front of the subarcuate fossa. Kermack and others (1981), however, incorrectly reinterpreted the lateral opening of the canalis pro-oticus in *Sinoconodon rigneyi* as being posterior to the subarcuate fossa (see also the Fig. 104 in *ibid.*).

In posterior aspect, after branching from the transverse sinus, the sigmoid sinus descended posteroventrally, and was contained in the groove of the sigmoid sinus at the posterior margin of the cranial surface of the petrosal. The sigmoid sinus passed by the aquaeductus vestibuli, and emerged through the jugular foramen to drain into the internal jugular vein. This restoration is quite similar to that of *Morganucodon* (see Kermack and others, 1981), which in turn is *Ornithorhynchus*-like.

However, Kielan-Jaworowska and others (1986) reconstructed the sigmoid sinus in *Nemegtbaatar* and *Chulsanbaatar* somewhat differently. They interpreted that only the dorsal segment of the sigmoid sinus became large, and the ventral segment drained into the vertebral vein through the foramen magnum, rather than draining through the jugular foramen into the internal jugular vein. They followed Butler (1967) in referring this condition of the absence of a sigmoid-jugular connection as monotreme-like (see also Fig. 12A in Butler, 1967). Incidentally, in this aspect as for many other cranial characters, there appears to be no such a thing as a "uniform monotreme condition." Lack of a sigmoid-jugular venous connection is characteristic only of *Tachyglossus* among monotremes. Therefore, if this presumed difference in ventral connection of the sigmoid sinus between *Lambdopsalis* and *Nemegtbaatar* plus *Chulsanbaatar* indeed existed, the former may be said as *Ornithorhynchus*-like and the latter *Tachyglossus*-like. Even so, the difference probably is taxonomically insignificant; Kielan-Jaworowska and others (1986) pointed out the extraordinary diversity in venous connections in both fossil and extant mammals. Moreover, Shindo (1915) pointed out that atrophy of the sigmoid sinus and the absence of an internal jugular vein are specializations in adult *Tachyglossus*; in the embryo of *Tachyglossus* both the sigmoid sinus and the internal jugular vein are present. This further trivializes phylogenetic implications of the difference.

The presumed difference in the ventral venous connection of the sigmoid sinus between *Lambdopsalis* and the Mongolian taxa may be more apparent than real. Posteroventral parts of the petrosal in *Nemegtbaatar* and *Chulsanbaatar* are poorly preserved, and "No specimen shows any more ventral segment of the sinus able to make a substantial connection with the region of the jugular foramen" (Kielan-Jaworowska and others, 1986, p. 560). In other words, absence of a sigmoid-jugular connection in *Nemegtbaatar* and *Chulsanbaatar* may be only a matter of interpretation, not necessarily supported by repeatable osteological observation.

Route of facial nerve.—Tracing the path of the facial nerve in early mammals has proved to be a difficult task due to the fragmentary nature of the materials; as a result, considerable misunderstanding of the associated structures has arisen (*cf.*, Kermack, 1963; Kielan-Jaworowska, 1971; Kermack and others, 1981; Crompton and Sun, 1985; and Kielan-Jaworowska and others, 1986). Restoration of the path of the facial nerve proper (the main trunk or hyomandibular branch) appears to be less a problem. Kermack and others (1981) incorrectly identified a fairly large foramen posterior to the foramen pseudovale in *Morganucodon* as the exit for the facial nerve proper; Crompton and Sun (1985), among others, subsequently pointed out that error. Nevertheless, interpretation of the route of the greater superficial petrosal nerve (the palatine branch of the facial nerve) has caused considerable confusion in understanding the structures surrounding the nerve.

Kielan-Jaworowska and others (1986, p. 585) defined the hiatus Fallopii as such: "in human: anterior orifice of canal carrying superficial petrosal nerve forwards from geniculate ganglion, is on endocranial aspect of tegmen tympani of petrosal. Tendency in palaeontology to describe 'hiatus Fallopii' on ventral aspect of cranium: probably a misnomer, though conceptually representing course of same nerve." Clearly, these authors realized the problem of probable misnomer of the structure, and yet considered the structure playing the same functional role. Consequently, they reconstructed the path of the greater superficial petrosal nerve in multituberculates in essentially the same manner as in previous literature. This practice, however, raises several problems:

1. As argued by Simpson (1937, p. 749), if restored as such, the "hiatus Fallopii" "certainly opens into the middle ear, and there is no evident probable way by which the superficial petrosal nerve, if issued here, reached the middle fossa of the skull or the sphenopalatine ganglion." I heartily agree with Simpson.

2. Only two possible relationships existed between the secondary side wall of the braincase and branches of the facial nerve: (1) the geniculate ganglion was internal to the secondary side wall and, after originating from the ganglion, both the greater superficial petrosal nerve and the facial nerve proper emerged through the foramen facialis secondarium before diverging in opposite directions; or (2) the geniculate ganglion was *extracranial*, and issued the two branches outside the braincase. Neither of these hypothetical conditions, however, finds its parallel in modern mammals.

3. According to De Beer (1937), the hiatus Fallopii in mammals is a gap between the prefacial commissure (an element of the primary side wall of the braincase) and the tegmen tympani (an element of the secondary wall), through which the greater superficial petrosal nerve passes. In other words, from its origin at the geniculate ganglion, the greater superficial petrosal nerve passes through its hiatus, runs anteriorly along the cranial surface of the petrosal beneath the trigeminal ganglion, and meets the deep petrosal nerve shortly before entering the Vidian canal (see also MacPhee, 1981). For much of its course, the greater superficial petrosal nerve is on the endocranial aspect of the tegmen tympani (Kielan-Jaworowska and others, 1986). However, if the condition in multituberculates was as these authors restored, the greater superficial petrosal nerve would have lain on

the extracranial aspect of the tegmen tympani and would have run along the tympanic surface of the petrosal. Such a condition, however, is extremely rare for mammals.

The well-preserved materials of *Lambdopsalis* render it possible to trace the path of the facial nerve with confidence and, as a result, resolve discrepancies between paleontological restorations and neontological observation. Based upon osteological features, the course of the facial nerve in *Lambdopsalis* is restored as follows: the facial nerve entered the well-defined internal auditory meatus, passed through the aquaeductus fallopii, and emerged through the foramen facialis primitivum behind the prefacial commissure. Then the facial nerve entered the cavum supracochleare, where the geniculate ganglion was located. The cavum supracochleare is immediately posterior to the semilunar fossa, and bounded medially by the prefacial commissure (representing the primary wall of the cranial cavity) and laterally by the tegmen tympani (representing the secondary wall). Within the cavum supracochleare, the geniculate ganglion gave rise to two branches. The greater superficial petrosal nerve passed anteroventrally, left the cavum epiptericum through the hiatus Fallopii, and then entered into the Vidian groove and continued into the pterygoid canal. The greater superficial petrosal nerve ran dorsomedially to the internal maxillary artery. The facial nerve proper (hyomandibular ramus) left the cavum supracochleare through the foramen facialis secondarium, and ran posteroventrally in the facial sulcus to emerge through the stylomastoid notch or foramen (the tertiary facial exit).

Interpretation of the route of the facial nerve in *Lambdopsalis* as represented above differs slightly from that presented by Kielan-Jaworowska and others (1986), but accords with the condition seen in modern mammals, especially in monotremes. The route is also similar to the path of the facial nerve in *Morganucodon* as reinterpreted by Crompton and Sun (1985, Fig. 6B). I suspect that the isolated petrosal of ?*Catopsalis joyneri* and the unidentified petrosals from the Hell Creek Formation described by Kielan-Jaworowska and others (1986) lack the more anteroventral part that would show the real hiatus Fallopii and Vidian groove. The greater superficial petrosal nerve must have left the cavum epiptericum more anteriorly than as restored by these authors. If, however, the true course was as these authors interpreted, the greater superficial petrosal nerve would have to have bent posteroventrally, left the cavum epiptericum behind the trigeminal ganglion (to enter the middle ear region), and passed through the "hiatus Fallopii" into the "?vidian groove" between the promontorium and the epitympanic recess. Finally, the nerve would have to have extended anterodorsally to enter the pterygoid canal. But this kind of loop-like path for the greater superficial petrosal nerve is quite problematic. It should be stressed that the "hiatus Fallopii" and "?vidian groove" of Kielan-Jawrowska and others (1986) probably are not equivalent to the structures with the same names as described herein. Their "hiatus Fallopii" is a notch at the posterior end of "?vidian groove," which more probably was vascular.

Expansion of vestibular apparatus.—The most striking and perplexing cranial feature of *Lambdopsalis* is its extraordinary, superficially bulla-like structure in the ear region. The original describers (Chow and Qi, 1978) considered the structure as a tympanic bulla, which led to the specific name of *L. bulla*. Kielan-Jaworowska and Sloan (1979) challenged Chow's and Qi's identification of the structure as real tympanic bulla, and suggested the pair of bulla-like structures as ". . . the inflated and probably somewhat ventrally displaced tabulars and paroccipital processes" (p. 195). Subsequently, however, it has been shown that the tabular bone was not present in multituberculates in general (Kielan-Jaworowska and others, 1984, 1986), nor was it present in *Lambdopsalis* in particular (this study). Kielan-Jaworowska and others (1986) modified the senior author's previous suggestion into: "In *Lambdopsalis* (Chow and Qi 1978) paroccipital process and mastoid strongly expanded to form 'tympanic bulla'" (p. 592). These authors further suggested that the expanded mastoid cavity of *Lambdopsalis* is comparable to the accessory air-spaces such as they described for the condylar cavity in some Mongolian multituberculates. They stated: "The existence of large air-sinuses in the middle ear of multituberculates may be compared with analogous adaptations in modern mammals (Lay, 1972)" (Kielan-Jaworowska and others, 1986, p. 557). However, all these suggestions were made prior to the time of opportunity for first-hand examination of specimens of *Lambdopsalis*.

As described above, the endocast of the hollow space within the bulla-like structure of *Lambdopsalis* represents the cast of the vestibule and semicircular canals of the *inner* ear. Besides, apertures such as the aquaeductus vestibuli, fenestrae vestibuli and cochleae open directly into the hollow space inside the inflated structure. These evidences suggest, beyond doubt, that the bulla-like structure is not part of the middle ear, and is rather part of the inner ear, the vestibular apparatus. In life, the space inside the structure was not air-filled, but rather housed the soft, fluid-filled vestibular organs.

Although rarely known in living mammals, expansion of the vestibular apparatus has been reported in members of other classes of tetrapods. In amphibians, especially the amphibian burrowers such as caecilians and ichthyophiids, the saccule is greatly expanded, and these animals were shown to be sensitive to low-frequency, substrate-borne vibrations and air-borne sound (Lewis and others, 1985; and Wever, 1985). Among reptiles, Bramble (1982) reported that gopher tortoises (especially the species *Gopherus polyphemus*) have a greatly expanded, otolithic inner ear, and suggested that it is functionally related to their burrowing habit. Although Bramble did not specify which parts of the inner ear in gopher tortoises are hypertrophied, it is known that the saccule is a more variable structure compared with the utricle and semicircular canals throughout the tetrapods (Lewis and others, 1985). It is also generally believed that the saccule in birds and mammals has become much diminished in size, and apparently has assumed a vestibular, rather than an acoustical, role (Lewis and others,

1985). However, exquisite acoustical acuity of the saccule was reported in squirrel monkeys (Young and others, 1976). Their study showed that the saccule is primarily responsive to gravitational and inertial stimuli, but some sensitivity to substrate vibrations is retained. Thus it is reasonable to suggest that expansion of the vestibular apparatus in *Lambdopsalis* may represent a similar adaptation to perception of low-frequency vibrations as seen in amphibian and reptilian burrowers. In their discussion of the middle ear specializations in *Lambdopsalis* (*e. g.*, the flat incudomalleal joint), Miao and Lillegraven (1986) suggested that *Lambdopsalis* was a burrower. The suggestion is strengthened by the structure of the inner ear discussed above; other evidence will be presented below.

In light of better understanding of auditory anatomy of *Lambdopsalis*, Kielan-Jaworowska and others' (1986) suggested analogy between multituberculates and some desert rodents requires reevaluation. In addition to Lay's (1972) study cited by Kielan-Jaworowska and others (1986), Webster and his associates have carried out thorough studies on the specialized auditory system in certain desert-dwelling rodents such as kangaroo rats, gerbils, and springhaas (Webster, 1961, 1962, 1970, 1973, 1977; Webster and Webster, 1971, 1972, 1975, 1977, 1980, 1984). They demonstrated that the rodents developed extraordinarily large middle ears to facilitate low-frequency sensitivity in detection of predators, including owls (with almost soundless flight) and rattlesnakes (with a brief, low-frequency sound at initiation of their strike). They also suggested that hypertrophied middle ears have evolved independently at least four times in rodent phylogeny. Similar auditory adaptation also was found in extinct South American argyrolagid marsupials (Simpson, 1970). I suggest that Kielan-Jaworowska and others' (1986) analogy between the middle ear of multituberculates and that of gerbilline rodents is no longer appropriate, simply because the hypertrophied bulla-like structure in *Lambdopsalis* has been demonstrated as part of the inner ear, rather than middle ear. The analogy holds only in a broad sense of their presumed similar functional demands for sensitivity to low-frequency vibrations. However, they achieved that sensitivity via quite different anatomical specializations. A detailed analysis of the functional morphology of the inner ear of *Lambdopsalis* is being developed for publication elsewhere.

Closely related to the expansion of the vestibule, the endolymphatic duct and sac of *Lambdopsalis* also became enlarged. These are inferred from the proportionately large size of the aquaeductus vestibuli and the recess of the endolymphatic sac. The endolymphatic duct is a tubular extension from the saccule which passes through the aquaeductus vestibuli and ends intradurally as the endolymphatic sac. In a strict sense, the endolymphatic duct is not a separately distinguishable duct, but is rather a constricted part of the saccule (Guild, 1927). It is not surprising, therefore, that *Lambdopsalis* possessed a greatly enlarged endolymphatic duct and sac resulting from the hypertrophy of its saccule. Besides, in a majority of mammals, the endolymphatic sac is usually flattened, lying between the layers of the dura mater, and only leaves a feeble (but perceptible) depression on the cranial surface of the petrosal (Bast and Anson, 1949). The recess of the endolymphatic sac in *Lambdopsalis* is, indeed, considerably more striking than might be expected.

Possible specializations in middle ear.—Auditory modifications also seem to have occurred in the middle ear of *Lambdopsalis*. In addition to possession of the flat incudomalleal joint (Miao and Lillegraven, 1986), *Lambdopsalis* appears either to have had a much reduced stapedial muscle, or to have lost it altogether; there is no perceptible fossa muscularis minor for attachment of this muscle.

The tensor tympani and stapedial muscles are the fundamental muscular components of the mammalian middle ear. As Fleischer (1978, p. 44) pointed out, "While nearly all mammals examined thus far have both muscles developed to a greater or lesser degree, nevertheless some exceptions were found. The stapedial muscle is missing in monotremes, but the tensor tympani is developed." The opposite is true in the pangolin (*ibid.*) and the eastern American mole (Henson, 1961). Both the stapedial and the tensor tympani muscles are very delicate in kangaroo rats (Webster and Webster, 1984), and the stapedial muscle is normally absent in a number of other fossorial rodents (Hinchcliffe and Pye, 1969; and Pye and Hinchcliffe, 1976). These intraaural muscles are also reduced in human ears (Fleischer, 1978). It is generally accepted that the reduction or loss of intraaural muscles is an adaptation to enhance low-frequency auditory acuity (Henson, 1961; Hinchcliffe and Pye, 1969; Fleischer, 1978; and Pickles, 1982).

In contrast, bats and dolphins have exceptionally well-developed intraaural muscles. According to Pye and Hinchcliffe (1976, p. 194-195), "Changes in middle ear muscle tonus modulate the frequency spectrum of external acoustic signals. The modulation provides a means whereby an air-borne auditory signal can be separated from a background noise. These modulations may also eliminate middle ear resonances without sacrificing the sensitivity of transmissions. Where the animal is concerned with ground-borne vibrations, this function of the middle ear muscles may be obviated." In other words, the middle ear muscles act to reduce transmission of low-frequency sounds, which can mask higher frequency stimuli over a wide range of frequencies. In fossorial mammals which gain information mainly through the low-frequency sounds, the intraaural muscles would be an impediment to hearing, and thus become greatly reduced or even lost.

Middle ear muscles, if present, usually leave recognizable depressions on the tympanic surface of the petrosal bone. Fossae muscularis major and minor were both described in ?*Catopsalis joyneri* by Kielan-Jaworowska and others (1986). There is also a well-defined fossa muscularis major in *Lambdopsalis*. Thus I suggest that absence of a well-defined fossa muscularis minor is significant, and indicative of reduction or loss of stapedial muscle. Coincidence of such characters as expansion of the vestibular apparatus, flat incudomalleal joint, the complete fusion of the cervical vertebrae, and reduction or possible loss of the stapedial muscle in *Lambdopsalis* are consistent with our functional interpretation (Miao and Lillegraven, 1986).

Promontorium and cochlea.—The promontorium (or promontory) of therian mammals generally refers to the rounded prominence within the tympanic cavity by outward projection of the first turn of the cochlea (MacPhee, 1981; Kielan-Jaworowska and others, 1986). Kielan-Jaworowska and others (1986, p. 595) pointed out, however, that the term "Has been loosely applied in palaeontology to ventral prominence of petrosal where cochlea or lagena is enclosed." The promontorium of *Lambdopsalis* is narrow, parallel-sided, relatively long, and houses a straight cochlea, in perfect accordance with Kielan-Jaworowska and others' (1986) description of the cochlea in Mongolian multituberculates. It is widely accepted that the promontorium is formed by part of the petrosal as a result of the underlying cochlea (Moore, 1981). However, Gow (1985) suggested that the mammalian promontorium is a homologue of the cynodont parasphenoid wing. His main argument is that the "parasphenoid" in *Tritylodon* seems to occupy a topographic position similar to that of the mammalian promontorium, and to be in sutural contact with the periotic, basisphenoid, and occipitals. If, however, the claimed suture between the "parasphenoid" and periotic in *Tritylodon* were in reality a crack, could not the bone in question be the part of the periotic rather than the parasphenoid? No matter whether the parasphenoid is retained (as I interpreted to be the case in *Lambdopsalis*) or lost unaccountably (as in many early mammals), the promontorium has been described as an inseparable part of the petrosal in early mammals. Without doubt, the promontorium in *Lambdopsalis* involves the anteroventral part of the petrosal, based upon clear sutural demarcations. Although Rowe (1986) chose to accept Gow's view, I find it particularly difficult to conceive that "The promontorium appears to have evolved first, in Mammaliaformes ancestrally (Table 8: chapter 4), and only afterwards did the lagena expand and coil" (Rowe, 1986, p. 198). Without even the slightest expansion of the lagena, how could the part of the bone have become prominent enough to deserve the name in the first place?

Rowe (1986) incorrectly suggested homology of the lagena of nonmammalian tetrapods with the mammalian cochlea, and used these terms interchangably. For instance, Rowe (1986, p. 198) stated: "In Monotremata and Theria, the lagena is greatly elongated, and has at least partial spiral curvature, forming the mammalian cochlea." The fact that monotremes have a cochlea with the lagena retained speaks for the lagena being only part of the greater cochlear structure (*i. e.,* distal end of the cochlear duct). In addition, the lagena is conventionally considered lost in therian mammals (Smith and Takasaka, 1971; Romer and Parsons, 1977; Wever, 1978; Fay and Popper, 1985; and Lewis and others, 1985). Admittedly, the use of the term "cochlea" as applied to nontherian mammals and other nonmammalian tetrapods is inappropriate, for only in the therians does the cochlea take on the snail-like appearance for which it was named. However, three major structural components are represented in the cochlea (*sensu lato*) of all tetrapods: the cochlear duct, perilymphatic duct, and basilar papilla (Romer and Parsons, 1977). As noted above, the lagena only represents the distal part of the cochlear duct in nontherian mammals and nonmammalian tetrapods, and is believed to be subsequently lost in therians. Thus the lagena cannot be equated with the mammalian cochlea in any sense. Nor can it be taken as an appropriate term for the whole cochlear structure in nontherian mammals and nonmammalian tetrapods. Therefore, the term cochlea should be kept in a broad sense to apply to the essentially homologous structure in reptiles and nontherian mammals, despite the morphological deviation from its etymology.

Yet another problem in Rowe's analysis of mammalian osteological characters merits comment. One of the most striking features of the modern therian auditory system is the coiled cochlea of the inner ear. By "coiled," I refer to a cochlea with more than one complete turn. Although without extended discussion, the coiled cochlea was treated as a synapomorphy of marsupials and eutherians by Miao and Lillegraven (1986). A straight or slightly curved cochlea in various early mammalian groups was regarded as plesiomorphic. The tendency toward coiling (that is, with curvature exceeding 180 degrees) in monotremes was considered intermediate, and coded as a monotreme synapomorphy in our cladogram (Miao and Lillegraven, 1986, Fig. 2). Rowe (1986) viewed the same set of data differently. He considered the cochlea with at least 180 degree curvature as a mammalian synapomorphy, and thus excluded morganucodontid triconodonts (with a straight cochlea) from the class Mammalia. As for multituberculates, Rowe (1986, p. 199) stated: "In light of the abundant data corroborating the placement of Multituberculata in Mammalia (se [sic] above), the presence of a straight cochlea, if confirmed, would represent a phylogenetic reversal."

By saying "if confirmed," Rowe seemed inclined to accept Sloan's (1979) interpretation of a hook-like cochlea in multituberculates. Prior to Rowe's writing, it had been documented, beyond reasonable doubt, that multituberculates possessed a straight cochlea (see Simpson, 1937; Hahn, 1978; and Clemens and Kielan-Jaworowska, 1979). Such has been reconfirmed by Kielan-Jaworowska and others (1986) and in the present study. Even if Sloan's (1979) claim were to be confirmed that *Ectypodus* had a cochlea spiraled 180 degrees, one still could argue that it may represent a derived condition within multituberculates. It remains conceivable that the straight cochlea of other multituberculates represents a primitive condition, contrary to Rowe's view. Moreover, if a straight cochlea in multituberculates should be deemed a phylogenetic reversal, one could equally argue that ". . . the lack of coiling in monotremes is the result of regression" (Fernandez and Schmidt, 1963, p. 158). However, this would violate the principle of parsimony, to which Rowe adhered.

Fernandez and Schmidt (1963) listed at least three reasons to support their suggestion that the coiled cochlea found in contemporary therian mammals evolved after the divergence from their common ancestors with those of monotremes:

1. Despite great variety in habits and habitats, no therian mammal has been shown to possess an uncoiled cochlea (see also Gray, 1955).

2. No evidence exists for cochlea coiling in advanced cynodonts or nontherian mammals other than monotremes.

3. A lagena is presumably absent only in therians, probably related to cochlear coiling. Fernandez and Schmidt (1963, p. 158) suggested: "If a coiled cochlea, without a lagena, had evolved before the monotreme-therian divergence, it is improbable that a typical lagena could have re-evolved in the monotremes."

Although we do not know the kind of cochlea possessed by eupantotheres and "tribotheres", it seems reasonable to suggest that the coiled cochlea represents a synapomorphy of marsupials plus eutherians, and the full coiling achieved after divergence between monotremes and the common ancestor of marsupials and eutherians. The curved cochlea of monotremes and the more or less straight cochlea of multituberculates reflect more primitive conditions. In this interpretation, early mammalian groups probably retained a more or less straight cochlea, just as obtained in their immediate reptilian ancestors.

Occipitals (Figs. 3, 10, 18, 28)

Description

The supraoccipital, paired exoccipitals, and basioccipital are fused completely, and thus treated together. These elements are well preserved in the following specimens: IVPP V5429, V5429.15, V7151.50, V7151.53, V7151.56-57, V7151.77, V7151.90, and V7151.95-96.

The basioccipital is best preserved in IVPP V5429 and V7151.56. In ventral view, the suture between the basioccipital and basisphenoid is at the level of the anterior tip of the promontorium, and approximately at a level medial to the foramen ovale inferium. Laterally, the basioccipital has a dorsal wing, which is in sutural contact with the petrosal; the suture is along the medial rim of the promontorium recess. Medial to the suture, and in the dorsal wing of the basioccipital, there is a recess, lying just opposite to the promontorium recess. Immediately posterior to these recesses, a large foramen is bounded laterally by the medial rim of the fenestra cochleae, medially by the posterior part of the dorsal wing of the basioccipital, and posteriorly by the occipital condyle. This foramen is designated as the jugular foramen. The occipital condyles protrude ventrally, forming crescent-shaped scrolls directed posterolaterally. In anteroventral view, each condyle bears a deep concavity facing anteriorly, here designated the condylar cavity. The dorsal wall of the condylar cavity plus the jugular foramen form a fairly broad area, here designated the jugular fossa. On the dorsal wall of the condylar cavity, a small foramen passes through the condyle, and opens into the cranial cavity. This foramen is designated the hypoglossal foramen.

In occipital view, there are no recognizable sutures between the basioccipital, exoccipitals, or supraoccipital. The supraoccipital and exoccipitals form a rough pentagon above the foramen magnum, with the top angle being the junction of the sagittal and lambdoidal crests. The foramen magnum is relatively small, and shaped as an elongated oval with its greater axis horizontally placed. The lesser (or vertical) axis of the foramen magnum is only about a quarter the total height of the occiput. The occipital condyles protrude behind the foramen magnum, and face downward and backward. The posterior margin of the basioccipital forms an indentation in the ventral edge of the foramen magnum between the anterior tips of the occipital condyles. Close to the suture between the petrosal and exoccipital plus basioccipital and at the dorsolateral corner of the occipital condyle is a small foramen, here designated the condylar foramen. The dorsal lobe of the occipital condyles does not reach the level of the top margin of the foramen magnum. There is no descending exoccipital process.

Discussion

The occipitals of *Lambdopsalis*, as those of mammals in general, are endochondral bones, and usually fused in the adult. They surround the foramen magnum, through which the brain communicated with the spinal cord.

The basioccipital is defined fairly well by sutures. The suture between the basioccipital and basisphenoid lies in a similar position to that restored by Kielan-Jaworowska and others (1986) for the Late Cretaceous Mongolian genera, in contrast to a more posterior position in *Ptilodus* reconstructed by Simpson (1937). In ventral view, the occipital condyles of *Lambdopsalis* are well separated, similar to those of *Nemegtbaatar*. The condyles are more separated than in *Chulsanbaatar* (see Kielan-Jaworowska and others, 1986). Occipital condyles of *Ectypodus* were restored as more widely separated (Sloan, 1979). The posterior margin of the basioccipital of *Lambdopsalis* is more indented than in the Mongolian genera. In other words, the posterior edge of the basioccipital of *Lambdopsalis* lies between the anterior tips of the occipital condyles, whereas the basioccipital of the Mongolian genera was restored as posteriorly wedging between the condyles nearly at a halfway point (Kielan-Jaworowska and others, 1986). The differences, however, may be more apparent than real, as posterior parts of the basicrania are generally broken in the Mongolian genera; restorations, as these authors noted, are tentative.

A striking feature in the basioccipital region of *Lambdopsalis* (and, to certain extent, of many Mongolian multituberculates) is the large size of the jugular foramen. This, however, may have resulted in part from exaggeration during preparation. Nevertheless, taeniolabidoid multituberculates seem to have had larger jugular foramina and fossae than plagiaulacoids and ptilodontoids (Kielan-Jaworowska and others, 1986). The jugular foramen presumably transmitted the glossopharyngeal (IX), vagus (X), and accessory (XI) nerves as well as the internal jugular vein. *Lambdopsalis* also possesses a single hypoglossal foramen, similar to *Ptilodus* (see Simpson, 1937) and *Ectypodus* (see Sloan, 1979); the foramen presumably functioned for passage of the hypoglossal nerve (XII). However, the hypoglossal foramen is double in *Paulchoffatia* (see Hahn, 1969), and is considered absent or indiscernible in the Mongolian taeniolabidoids (see Kielan-Jaworowska, 1971, 1974; Kielan-Jaworowska and others, 1986). As there is only a thin bony bridge separating the hypoglossal foramen from the jugular foramen in *Lambdopsalis*, I suspect that, if originally present, even a thinner bony partition between the foramina in many smaller Mongolian genera would be more vulnerable to breakage during preparation.

Another striking feature of the occipital region in *Lambdopsalis* is presence of a large, hollow concavity immediately posterior to, but confluent with, the jugular fossa. It was also described in many of the Mongolian taxa, notably in *Chulsanbaatar* and *Kryptobaatar* (see Kielan-Jaworowska and others, 1986). This structure has been termed the condylar cavity, which is adopted here for the similar structure in *Lambdopsalis*. Kielan-Jaworowska and others (1986) also described a bony partition present between the jugular fossa and the condylar cavity, and called it the lateral wall of the condylar cavity or the medial wall of the jugular fossa. However, I have been able to recognize such a bony partition neither in *Lambdopsalis* nor in the Mongolian specimens (*personal observations*).

Furthermore, Kielan-Jaworowska and others (1986) suggested the condylar cavity served as accessory air-space, ". . . comparable with the expanded mastoid cavity of the Paleocene *Lambdopsalis*" (p. 557). As noted in the preceding section, however, they incorrectly made analogy between these presumed air-spaces in multituberculates with expansion of the middle ear cavity in some desert-dwelling rodents.

LOWER JAW (Figs. 4, 26, 31, 32)

Description

The lower jaws of *Lambdopsalis* consist of only a single paired bone, the dentary. The dentary is well preserved in the following specimens: IVPP V5429, V5429.10-11, V7151.50, V7151.90, and V.7152.65. Description of the dentary is supplemented by observation of the following, partially preserved specimens: V5429.17-21, V7152.17, V7152.39, V7152.48, V7152.50, and V7152.66.

The most complete specimen of the dentary bone is the left mandible of the holotype (V5429). The animal represented, however, was a juvenile, with the second molar unerupted. The following description is based mainly upon the left dentary of specimen V5429. Wherever differences arising from allometric growth exist, both juvenile and adult conditions will be described.

The dentary is rodent-like in general appearance. It is strong and massive. Anterior to the premolar the bone is transversely-compressed and sheath-like, encasing the strongly procumbent lower incisor. The dorsal margin of that part of the dentary floors the diastema, and looks like the bottom part of a U-shaped tube. Beneath the bottom of the U-shaped diastema is a small mental foramen. The ventral margin of the dentary forms a smooth concave curve from the incisor alveolus to a point below p4, where it flattens out posteriorly to become roughly straight. The alveolar border is slightly inclined medially. An ascending ramus arose at a point beneath the middle of m1 in the juvenile, and at a point beneath the anterior edge of m2 in the adult. The ascending ramus is designated as the coronoid process. There is a wide, shallow trough between the alveolar border and the coronoid process. The coronoid process leans laterally, and is roughly triangular and blade-like, with a thin and almost vertical posterior edge.

In lateral view, there is a large, moderately deep fossa below the coronoid process, the masseteric fossa. The masseteric fossa is delimited anteriorly by a distinct ridge descending from the anterior edge of the coronoid process. The ridge makes a U-turn at a point beneath p4, and then extends posteriorly to form an almost straight ventral margin of the masseteric fossa. Posterior to the masseteric fossa, the dentary gives rise to a slightly elevated process. The process becomes bulged at its posterior end into the mandibular condyle. In posterior view, the condyle is oval-shaped, with its greater axis dorsomedially-ventrolaterally oriented.

In medial view, posteroventral to m2, a large and deep fossa, the pterygoid fossa, faces posteromedially. The pterygoid fossa is roofed by the alveolar border, and floored by a wide bony shelf inflected from the posteroventral part of the dentary. The bony shelf gradually diminishes posteriorly. The shelf is designated the angular process. It cannot be seen in lateral view. The pterygoid fossa was deeper in the juvenile than in the adult. In the anteroventral corner of the pterygoid fossa a single, large foramen opens into the inferior dental canal, here designated as the inferior dental foramen. Beneath the alveolar border, the medial surface of the dentary forms a broadly concave area. The surface of the symphysis is slightly rugose, and the two dentary bones in the lower jaw appear not to have fused together.

Discussion

The dentary of *Lambdopsalis* is so characteristic of multituberculates that little needs to be added to the existing discussions (*e. g.*, Broom, 1914; Simpson, 1928, 1937; Granger and Simpson, 1929; Jepsen, 1940; Clemens, 1963; Kielan-Jaworowska, 1971, 1974; Clemens and Kielan-Jaworowska, 1979; and Sloan, 1979). Except a few early authors (*e. g.*, Cope, 1881, 1882; and Falconer, 1857), however, most workers have denied (either explicitly or implicitly) the presence of an angular process in multituberculates. Although Matthew and Granger (1925) described an inflected angular process in *Prionessus*, subsequent authors, although agreeing that there exists a prominent, inflected flange along the medial side of the posteroventral border of the dentary in multituberculates, referred to it by different names. The flange was variously called the "pterygoid crest" (Simpson, 1926, 1928, 1937; Granger and Simpson, 1929), "shelf of the

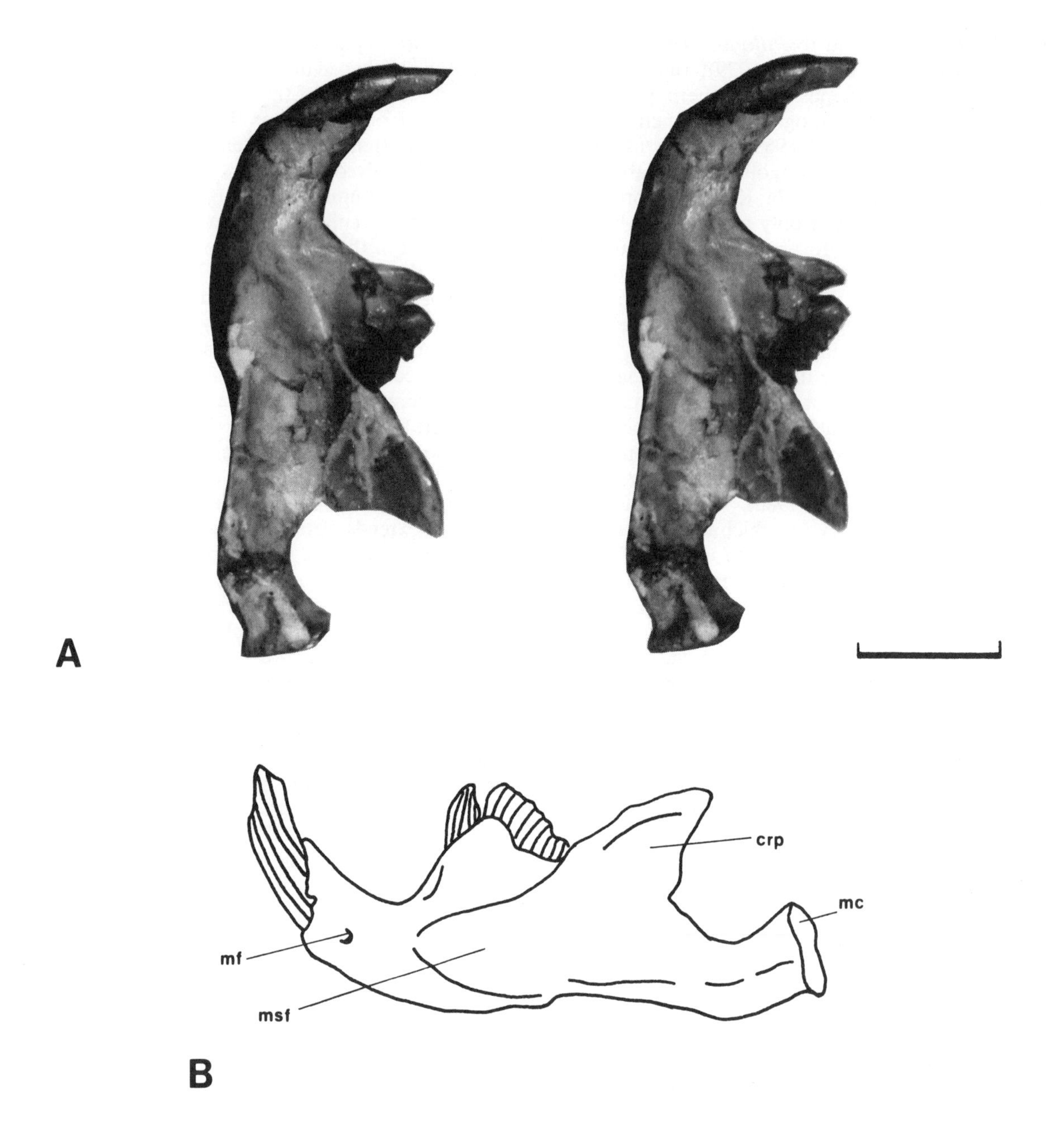

Figure 31. ***Lambdopsalis bulla*****, V5429, holotype; lateral view of left mandible. *A*, stereophotograph, anterior to top; and *B*, explanatory drawing, anterior to left: *crp*, coronoid process; *mc*, mandibluar condyle; *mf*, mental foramen; and *msf*, masseteric fossa.**

pterygoid fossa" (Jepsen, 1940), "floor of the pterygoid fossa" (Clemens, 1963), and "the wide, horizontal floor of the pterygoid fossa" (Kielan-Jaworowska, 1971, 1974).

As to the function of the flange-like structure in multituberculates, all authors considered it as the point of insertion of the internal pterygoid muscle. Thus it is clear that both functionally and topographically the flange-like structure of multituberculates is similar to the angular process of other mammals. Curiously, however, Broom (1914) regarded them as homologous structures, although he nevertheless protested earlier authors' designation of the flange-like structure in multituberculates as a true angular process.

Adams (1919) showed that jaw muscles of mammals are remarkably constant throughout the group, except in monotremes and edentates in which the reduced or modified condition of the dentition and mandible makes the pterygoid muscles of little functional importance. Based upon similarities in general appearance of the dentition and lower jaw, Simpson (1926) restored the jaw musculature of multituberculates along the pattern of modern rodents. He (1926, p. 236) suggested that, in multituberculates, "The pterygoid muscles were remarkably developed, as is shown by the high and long pterygoid crest along the inner side of the postero-inferior mandibular border. The function of the pterygoid muscles is to

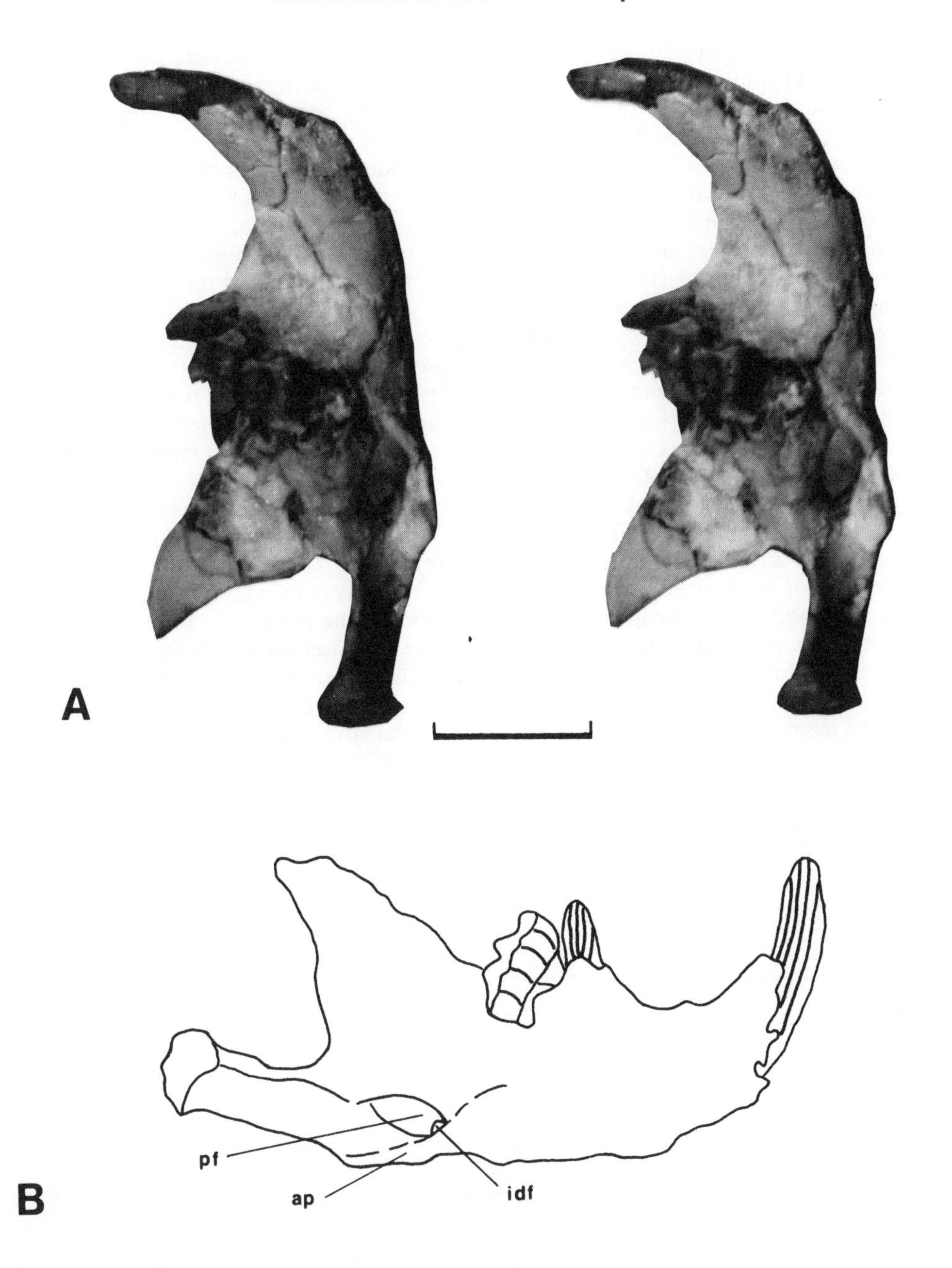

Figure 32. ***Lambdopsalis bulla*, V5429, holotype; medial view of left mandible. *A*, stereophotograph, anterior to top; and *B*, explanatory drawing, anterior to right: *ap*, angular process; *idf*, infradental foramen; and *pf*, pterygoid fossa.**

help close the jaws, to steady the lower jaw against the outward pull of the temporal and masseter muscles, and especially to pull the jaw forward and make possible the grinding motion of the teeth in herbivorous forms." To counter the then-prevailing belief of monotremes' affinity with multituberculates, Simpson (*ibid.*) noted that there is no particular reason to assume that multituberculates exhibited monotreme peculiarities in jaw musculature. Thus, to Simpson, there was no doubt that the well-developed pterygoid crest in multituberculates was adapted to development of strong pterygoid muscles, functionally just as in other herbivores. However, Simpson was unable to determine whether there existed one or two insertions for the pterygoid muscles. Simpson's view of the restoration of multituberculate jaw musculature was essentially shared by Turnbull (1970). Although the evidence is unclear, Sloan (1979) believed that only a single pterygoid muscle existed in *Ectypodus*.

Although Simpson restored the jaw musculature of multituberculates based upon that of modern rodents, and admitted existence of a well-developed functional correlate of the mammalian angular process (*i. e.*, the pterygoid crest), he did not regard the pterygoid crest of multituberculates as a true angular process. Perhaps this reluctance can be understood in the context of Broom's (1914) original argument. Broom was concerned by the situation that "By the large majority of writers the 'inflected angle' of the Multituberculates has been brought forward as a Marsupial character" (p. 129). In fact, multituberculates once were considered to be marsupials (*e. g.,* Cope, 1884; and Gidley, 1909). Broom, who believed in monotreme affinity of multituberculates, naturally saw more dissimilarities in morphology of the inflected angle between multituberculates and marsupials. Simpson (1926) and Granger and Simpson (1929) stressed that the inflected angle of multituberculates did not extend to the very posteroventral end of the dentary, and thus deemed that it is ". . . in no way resembling an angular process save in function" (Granger and Simpson, 1929, p. 612). Broom (1914) said about this: "There is no doubt that the borders are homologous but owing to the extremely backward position of the articulation in the Multituberculates the internal pterygoid muscle has to be inserted further forward on the jaw" (p. 130).

It is now known that an inflected angular process is not a unique characteristics of marsupials. It also occurs in some modern insectivores, and Mesozoic eupantotheres and eutherians. Thus it seems reasonable to admit that multituberculates have a true inflected angular process. There exists great variation in detailed morphology of the angular process in modern mammals. The angular process is not necessarily restricted to being a projection from the very posteroventral end of the dentary, contrary to Simpson's suggestion. As Ride (1957) noted, similarly inturned pterygoid shelves also may occur in mandibles of modern eutherians along the posteroventral border of the angular process. Ride (1957, p. 400) also pointed out, in multituberculates, ". . . the angle is not inflected in the usual manner of marsupial herbivores in which the floor of the pterygoid fossa is triangular in surface with the apex directed anteriorly."

The flange-like structure in the dentary of *Lambdopsalis* is so well-developed that it can reasonably be called nothing but the angular process. From the above discussion, it is clear that historically the term "angular process" was abandoned in multituberculates in just the way the baby was thrown out with the bath water. One can show, through this example, how easily we can be constrained by intellectual heritage, or sometimes even prejudice.

Finally, I interpret, as seen invariably among mammals: (1) the masseteric fossa in *Lambdopsalis* was the place for insertion of the masseter muscle; and (2) the inferior dental foramen and the mental foramen served as the entrance and the exit, respectively, for the inferior alveolar artery and nerve.

SECTION III

GENERAL DISCUSSION

Invalidity of Lambdopsalidae

Lambdopsalis bulla was named by Chow and Qi (1978) and assigned to a new monotypic family, Lambdopsalidae. Its taxonomic validity at all three levels was challenged immediately by Kielan-Jaworowska and Sloan (1979). Subsequently, examination of new materials of *Lambdopsalis bulla* from the Bayan Ulan locality led to the conclusion that both the genus and species are valid (Miao, 1986). However, the problem of validity of Lambdopsalidae was left open, and evidence conducive to a solution was expected to emerge from the study of the cranial anatomy of *Lambdopsalis bulla*. In the meanwhile, however, various authors have treated the problem based upon their own judgments, usually without further comments. For example, Lambdopsalidae was considered valid by Hahn (1983) and by Li and Ting (1983), but invalid and as a junior synonym of Taeniolabididae Granger and Simpson 1929 by Hahn and Hahn (1983) and by Simmons and Miao (1986).

As noted earlier in the section entitled "Petrosal and Auditory Region," expansion of the vestibular apparatus in *Lambdopsalis* was mistakenly identified as an inflated tympanic bulla by the original describers (Chow and Qi, 1978). They (*ibid.*, p. 85) stated: ". . . the new form is singularly characterized by the presence of a pair of fully developed tympanic bulla of enormous size, a character which is quite unique and renders it difficult to affiliate it to any known multituberculate group." This was their primary evidence in supporting a separate familial assignment for *Lambdopsalis*. In addition, they also were impressed by much more well-developed shelf-like flange in the posterior part of the medial side of the dentary (Chow, *personal communication*).

As the present study has shown, neither of these characters is unique to *Lambdopsalis*. A similarly expanded vestibular apparatus existed in ?*Catopsalis joyneri* (*personal communication* from Luo, and my observations), and probably in *Kamptobaatar* as well (*personal observations*). A shelf-like flange identified here as the angular process has been reported in almost all multituberculates. Both of these features are more strongly developed in *Lambdopsalis*. Except for differences in degree, however, there is nothing especially peculiar about *Lambdopsalis* in those cranial characters. To the contrary, most dental and cranial characters show phylogenetic unity of *Lambdopsalis* with the other taeniolabidid genera (see also Miao, 1986; Simmons and Miao, 1986). Therefore, I conclude that the Lambdopsalidae Chow and Qi, 1978 is a subjective junior synonym of the Taeniolabididae Granger and Simpson, 1929.

Biogeographic and Ethological Implications

The assessment of *Lambdopsalis* as a member of Taeniolabididae as based upon cranial characters strengthens results of phylogenetic analyses of taeniolabidids derived from dental features. Cranial evidence indicates that *Lambdopsalis* could have been derived from a *Catopsalis*-like creature. This accords well with analyses of dental characters (Miao, 1986; and Simmons and Miao, 1986). Accordingly, the present study supports the thesis that ancestors of the *Sphenopsalis* + *Prionessus* + *Lambdopsalis* lineage originated in North America (probably from, or shared the latest common ancestor with, *Catopsalis*) and then dispersed from North America back to Asia (see also Simmons and Miao, 1986). However, this hypothesis may be subject to more rigorous tests by future discovery of multituberculates in lower horizons of Asian Paleocene rocks.

As discussed earlier in this paper, substantial evidence derived from cranial anatomy suggests that *Lambdopsalis* was a burrower. It is possible that ?*Catopsalis joyneri* had similar adaptations, as indicated by its expanded inner ear. This would require some modification of Jenkins and Krause's (1983) generalization about the strikingly homogeneous locomotor adaptations for climbing presumed to exist among North American multituberculates. In contrast, the present study led me to suspect that multituberculates may have been nearly as diverse in their behavior and locomotion as modern rodents. This is also indicated by high variation in body size, great diversity in dental specializations, long phylogenetic history, great taxonomic diversity, and rich fossil record of multituberculates (Clemens and Kielan-Jaworowska, 1979; Sloan, 1979). Multituberculates probably had more diverse locomotor specializations and occupied a broader spectrum of ecological niches than suggested by the limited postcranial remains available to Jenkins and Krause (1983).

Furthermore, although Jenkins and Krause's (1983) hypothesis about the arboreal adaptation of *Ptilodus* at first seems convincing, several problems may deserve consideration. First, the activities of animals are constrained not only by morphological specializations but also by behavioral factors. For instance, gray foxes climb trees whereas red foxes do not. One cannot learn by dissection which species climbs or, indeed, that either climbs. Also, even without special structural modifications in adaptation to a particular kind of locomotion, many animals are capable of jumping, swimming, climbing, running, and digging. Contrariwise, some animals may locomote contrary to ways in which they seem to be morphologically adapted. For example, adult gorillas have the structures that usually correlate with arboreal brachiation; simply because of large size, however, they seldom do so (Hildebrand, 1974).

Second, we often look at animals quite anthropocentrically. For our eyes, for example, sagebrushes could well have appeared as forests to shrew-sized multituberculates. Thus, what may seem terrestrial to us (because they lived in low ground-cover) may actually have been arboreal, involving climbing, gripping, and spanning discontinuous substrata. Because of small size, however, they may have had morphological specializations in their skeletons generally associated with arboreal life.

In summary, the present study suggests that, in addition to the arboreal type, at least some multituberculates probably were fossorial. Further, the presumed homogeneity in locomotor specializations may not reflect the whole diversity enjoyed by multituberculates either in habits or in habitats.

PHYLOGENETIC POSITION OF MULTITUBERCULATA

Introduction

Simpson (1928, p. 163) wrote: "It is probable that no other group of animals has given rise to such wide diversity of opinion regarding its relationships as has the order Multituberculata. They have been referred to each of the known mammalian sub-classes and to a distinct sub-class of their own. As to their more intimate relationships, at least ten or twelve mutually exclusive views have been expressed and warmly defended— this despite the fact that they are now rather well known, much better than any other group of Mesozoic mammals."

Collections of multituberculates, especially their cranial remains, have greatly expanded both geographically and chronologically since the time of Simpson's writing; yet his frustration remains with us (and, to certain extent, has become even stronger). Not only does their affinity remain uncertain, but even their unity as a natural group has been questioned (Rowe, 1986). However, one can be more optimistic about progress simply by observing that several paths toward the improbable have been closed (*e. g.*, multituberculate affinities with marsupials or rodents). Fewer reasonable alternatives exist now than when Simpson (1928) wrote. In the following discussion, I will first defend monophyly of multituberculates, and then consider the relevance of the present study to affinities of multituberculates. I arbitrarily restrict Multituberculata to include only non-haramiyid multituberculates. Although Hahn (1973) suggested that haramiyids were probably members of the multituberculate stock, Clemens and Kielan-Jaworowska (1979) treated haramiyids as "?Multituberculata *incertae sedis*" due to what Simpson called "painful uncertainty." Limited information of dentitions of haramiyids contributes little to the present discussion, and thus will be ignored.

Monophyly of Multituberculata

Rowe (1986) expressed concern that monophyly of the Multituberculata may be threatened by the traditional inclusion of plagiaulacoids as primitive members of the order. He stated: "For this reason, in the following discussion and in accompanying figures (*e.g.*, Fig. 4) I restrict the name Multituberculata to include only Taeniolabidoidea and Ptilodontoidea, and I treat *Kuhneodon* [sic] and *Paulchoffatia* separately" (p. 143). Besides the two genera that Rowe mentioned, Plagiaulacoidea includes a variety of genera, which together represent a fairly diverse early radiation of multituberculates.

I consider Rowe's concern to be superfluous. As Clemens and Kielan-Jaworowska (1979, p. 136) summarized, "Dental specializations of a single pair of procumbent lower incisors, blade-like lower premolars, multicusped molars, and evolution of a pattern of mastication involving orthal and propalinal but not transverse mandibular motion speak to the unity of the order. These characters also unambiguously isolate multituberculates from other nontherians." Other dental peculiarities that can be added to the above listings include: (1) never more than two molars; (2) enlargement of the second upper incisor; and (3) reduction in the midst of the premolar series, rather than at its ends. Also among the dental characters, "The pattern of occlusion of the molars has appeared to be a firm, diagnostic character of the Multituberculata" (Clemens and Kielan-Jaworowska, 1979, p. 113). In all known multituberculates thus far, m1 and m2 are aligned whereas M1 and M2 are laterally offset. Although Hahn (1969) reported that the M2 in *Paulchoffatia* and *Pseudobolodon* is aligned directly behind M1 rather than offset lingually relative to M1, his interpretation was questioned subsequently by Clemens and Kielan-Jaworowska (1979). Recently, Hopson confirmed that the occlusive pattern of molars in *Paulchoffatia* and in *Pseudobolodon* is indeed characteristic of multituberculates, contrary to Hahn's original interpretation. Therefore, this unique occlusal pattern of the molars in all known multituberculates appears autapomorphic, strongly supporting the notion of monophyly of the order Multituberculata.

The cranial characters provide added support for the notion of monophyly of multituberculates. Among the three suborders certainly referable to the order, their skull morphology, especially in the ear region, shows striking similarities, and is distinguishable from that of other mammalian groups. This was summarized thoroughly by Clemens and Kielan-Jaworowska (1979), and also has been borne out by more recent studies (Hahn, 1981; Kielan-Jaworowska and others, 1986; and this study). Although few of the cranial features are unique, the observed combination of characters seems distinctive. The characteristic combination includes: (1) a partially roofed but unfloored orbit; (2) a strong zygomatic arch with a slender jugal; (3) firm establishment of the secondary side wall of the braincase, with little reduction of the primary one; (4) acquisition of a triossicular middle ear accompanied by a straight cochlea; (5) a typical mammalian palate coupled with an abbreviated post-palatal region; and (6) a mandible with its articular condyle extremely backwardly placed, and with a unique shelf-like angular process.

Although Krause and Jenkins (1983) documented a number of unusual (and even unique) postcranial features in ptilodontoid and taeniolabidoid multituberculates, these have little bearing on particular concerns under discussion. We do not yet know if those postcranial features also characterize plagiaulacoids. Nevertheless, even in the present weak state of knowledge, it seems beyond reasonable doubt that the order Multituberculata, as traditionally defined, is monophyletic. Rowe's exclusion of plagiaulacoids from multituberculates is without factual support.

Phylogenetic Relationship Between Multituberculata and Living Theria

After splitting the traditionally defined order Multituberculata into *Paulchoffatia* plus *Kuehneodon*, and "Multituberculata," Rowe (1986) further suggested that the plagiaulacoids should be placed ". . . in the lineage that includes living Theria" (p. 224). He conceived that multituberculates are closer to Theria than to Monotremata. He also considered that multituberculates are closer to living Theria than either monotremes or eupantotheres are (see Rowe, 1986, Fig. 4).

The sole evidence upon which this systematic departure was based is that multituberculates ". . . share one derived character state with Theria, a reduction in the number of exits for the infraorbital canal onto the face (Hahn, 1985)" (Rowe, 1986, p. 224). As already discussed in the section entitled "Maxilla," that character is too variable to be meaningful in phylogenetic analysis, let alone to be used as a character of such importance in the present evolutionary issue. Even if that character were valid, "It has been the experience of zoologists that major groupings of animals on single characters are seldom reliable" (Parrington, 1971, p. 267). In addition, by accepting Rowe's scheme of revision, one must assume that major evolutionary reversals had occurred. I suspect, therefore, that the close relationship between multituberculates and living therians as suggested by Rowe is unlikely.

Phylogenetic Relationship Between Multituberculata and Monotremata

Close relationship between multituberculates and monotremes has been advocated on several occasions (*e. g.*, Cope, 1888; Broom, 1914; Winge, 1923; Kermack and Kielan-Jaworowska, 1971; Kielan-Jaworowska, 1971; and Kermack and Kermack, 1984), and has been rejected almost as often (*e. g.*, Simpson, 1928, 1937; Granger and Simpson, 1929). Cope's (1888) original proposal of monotreme affinity with multituberculates was based upon presumed dental similarities between *Ornithorhynchus* and multituberculates. Simpson (1929), however, demonstrated that "The teeth are quite dissimilar, the cusps differing fundamentally in number, arrangement, form, and function" (p. 12). Simpson went on: "On the basis of the teeth alone it would be as reasonable to relate *Ornithorhynchus* to *Homo* as to multituberculates" (*ibid.*, p. 12). Broom's (1914) view was developed on the basis of cranial features. However, Simpson (1928, 1937) and Granger and Simpson (1929) pointed out that the observed resemblences between multituberculates and monotremes were either superficial or symplesiomorphic, neither of which is indicative of affinity. More recent revival of the hypothesis (Kermack and Kielan-Jaworowska, 1971) has received only mild criticism (*e. g.*, Griffiths, 1978) and hesitant support (*e. g.*, Kemp, 1982, 1983).

Griffiths (1978) opposed Kermack and Kielan-Jaworowska (1971) on the basis of several varieties of inaccuracies and ambiguities included in their arguments for close relationship between multituberculates and monotremes. Griffiths (1978, p. 331) decided: "However these are but minor matters; Kermack and Kielan-Jaworowska have brought exciting new light to bear on the theses of Cope and of Broom." Kemp (1983) realized that the few similarities that multituberculates share with monotremes may, indeed, indicate a close relationship between the two groups. However, Kemp (1983, p. 353) qualified his assessment by stating: "The multituberculates may be related to the monotremes." Kemp further expressed his hesitation in his cladogram of phylogenetic relationships of mammalian groups by using a dashed line plus question mark to place Multituberculata as possible, but uncertain, sister group of Monotremata (Kemp, 1983, Fig. 15). As is often the case, hesitation and uncertainty by the original researchers became diluted (or even disappeared) in secondary sources or in standard textbooks (*e. g.*, compare Kemp, 1983, Fig. 15 with Schoch, 1984, Table 1, and with Tyndale-Biscoe and Renfree, 1987, Fig. 10.1). Even though one of its more recent proponents no longer holds the thesis (Kielan-Jaworowska, 1983; and *personal communication*), it would be fair to say that the idea of monotremes' affinity with multituberculates remains as a possible alternative. Therefore, it is relevant to reevaluate validity of the hypothesis.

Kermack and Kielan-Jaworowska (1971, p. 103) suggested that: "Multituberculata and Monotremata have in common: (1) greatly reduced alisphenoid, (2) jugal absent, (3) lacrimal usually reduced or absent, (4) multituberculate ectopterygoid resembling the monotreme 'Echidna pterygoid.' Multituberculata and Monotremata are clearly related." However, none of these characters still holds. Great reduction of the alisphenoid is observed in monotremes, but may not be true of multituberculates in light of the information derived from the present study. Presence of a jugal has been reported in *Ptilodus* (*personal communication* from Hopson) and in a number of Mongolian multituberculate genera (Kielan-Jaworowska and others, 1986). Among monotremes, the jugal is present in *Ornithorhynchus* but absent in *Tachyglossus*. The lacrimal may be absent in certain genera of multituberculates (*e. g., Lambdopsalis*) but is moderately large in *Paulchoffatia* (see Hahn, 1969) and very large in the Mongolian genera (Clemens and Kielan-Jaworowska, 1979). Reported absence of the lacrimal in the Mongolian taxa (Kielan-Jaworowska, 1971) was corrected subsequently by the same author (1974). As discussed above, multituberculates may not possess an ectopterygoid bone. Besides, as Griffiths (1978, p. 330) stressed, ". . . the ectopterygoids in echidnas and platypuses do not resemble one another." Subsequent studies also showed that both the postcranial skeleton (Clemens, 1979) and the endocranial casts (Kielan-Jaworowska, 1983) of multituberculates suggest major differences from monotremes.

After pondering whether Kermack and Kielan-Jaworowska's (1971) view was credible, Griffiths (1978, p. 331) wrote: "If it turns out that the multituberculates had three ear ossicles it may indicate parallelism or perhaps relationship but if it were found that the incus was ankylosed to the malleus and that the latter had a long gracile process cemented to the tympanic it would, to an overt incorrigible typologist like myself, strengthen the argument that multituberculates and monotremes are related." Most of these conditions were fulfilled by the discovery of the three ear ossicles in *Lambdopsalis* (see Miao and Lillegraven, 1986), yet we remained in a helpless position in deciding degree of relationship between multituberculates and monotremes. Surprisingly, no matter whether the three ear ossicles were assumed to have evolved only once or more than once in mammalian phylogeny, none of the constructed alternative cladograms resulted in close proximity between multituberculates and monotremes (see Miao and Lillegraven, 1986, Fig. 2).

Archer and others (1985) reported discovery of an Early Cretaceous Australian monotreme (*Steropodon galmani*) with tribosphenic molars. On the basis of dental specialization, they regarded monotremes as therian mammals. More recently, however, Kielan-Jaworowska and others (1987) challenged the tribosphenic nature of the molars of *Steropodon*. Instead, they deemed that Monotremata ". . . appears to have been derived from therians before the development of tribosphenic teeth, possibly during the Jurassic period" (p. 871). Although one still could argue the possibility that multituberculate molars also were derived from the therian lineage, the basis of Cope's (1888) initial assertion of a close relationship between multituberculates and monotremes certainly collapsed upon Archer's and others' (1985) discovery. After a century-long debate, Cope's idea of close relationship between multituberculates and monotremes should be abandoned.

Problem of Nontherian/Therian Dichotomy

Despite uncertainties of the more intimate relationships of Multituberculata, the prevalent view of paleomammalogists has been to place Multituberculata within the subclass Prototheria (*e. g.*, Hopson, 1970). The class Mammalia was believed to include two fundamentally split lineages, Prototheria and Theria. As discussed earlier in the section entitled "Alisphenoid," this idea has been seriously questioned more recently on the basis of developmental studies. Kemp (1983) forcefully demonstrated that no particular synapomorphy exists that could satisfactorily define Prototheria as a monophyletic group.

As Kemp (1983) ably summarized, only two characters have been used to define nontherians as the sister group of therians, namely the molar and braincase structures. Various authors (*e. g.*, Simpson, 1929; Hopson and Crompton, 1969; and Kemp, 1983) have pointed out that the teeth of various nontherian orders are not really comparable to one another. The primary argument for a nontherian/therian dichotomy actually was based upon the presumed difference in braincase structure. Understandably, once Presley's (1981) hypothesis gained acceptance, the dichotomy seemed to most researchers to break down (exceptions include Kermack and Kermack, 1984). If, indeed, Kermack and others' (1981) claim holds true concerning the cranial structure in nontherians, the problem of an early mammalian dichotomy still may be taxonomically real. Griffiths' (1978) and Presley's (1981) embryological studies can be used to argue for a monophyletic origin of mammals, but they may not be used to refute a dichotomous separation of prototherians versus therians in taxonomy. In other words, the presumed difference in braincase structure, if it existed, still would carry some taxonomic significance, even though less fundamental than once thought. After all, no other cranial character of taxonomic value has been evaluated similarly in terms of its ontogeny and phylogeny. In fact, as one of the advocates of the hypothesis, Hopson (1964) was fully aware that the so-called anterior lamina is actually an intramembranous ossification. Therefore, for those who still believe in reality of the dichotomy, an expectation would be to discover a pre-Cretaceous therian with a "typical" therian braincase. But according to Kemp (1983, p. 374), "It is perfectly possible for *Kuehneotherium* and the eupantotheres to have possessed a morganucodontid-like braincase." Speculation about the yet-to-be-known cannot be used to convince either side of the debate.

So far no one seems to have doubted the reality of a uniform nontherian type braincase pattern. However, as discussed earlier in the section "Alisphenoid," *Lambdopsalis* possessed true foramina ovale inferium and masticatorium perforating the expanded alisphenoid. According to Popper's Principle of Falsification, this discovery may immediately break down the demarcation of nontherian/therian dichotomy as based upon the braincase structure. Since this discovery, I have been haunted by the following question: Is there any solid paleontological evidence which demonstrates clearly that an anterior lamina perforated by trigeminal foramina really existed in multituberculates, docodonts, triconodonts, or monotremes? To my surprise, it turned out that evidence for a negative answer to the question is overwhelming. Osborn's (1895, p. 436) statement seems especially pertinent: "We may waive our applications of these facts to theories, but let us not turn our backs to the facts themselves."

1. Monotremes are said to have an anterior lamina, though embryologically it is known not to be an outgrowth from the periotic (Griffiths, 1978; and Presley, 1981). In any case, the mandibular branch of the trigeminal nerve does not pass through the petrosal in monotremes (Griffiths, 1978).

2. The only available cranial material of docodonts is a dorsoventrally crushed skull of *Haldanodon*, yet to be studied, in the collection of Institute of Paleontology, Free University in West Berlin. Although both cranial roof and basicranium of *Haldanodon* are well preserved, the orbitotemporal region cannot be observed due to crushing (*personal observations*). However, concerning the anterior lamina, Kermack and others (1981, p. 136) suggested: "it was certainly present in the Docodonta as well, since it is found in the other two suborders—Morganucodonta and Eutriconodonta, which make up the order Triconodonta." Circular reasoning has been substituted for observation.

3. Jenkins and Crompton (1979) once commented on the condition in triconodonts as such: "It should be noted that the materials available to Kermack were dissociated and, in some cases, damaged. However astute his observations were, it is now certain that a much clearer understanding of triconodont braincase structure can be achieved with reference to the recently discovered skulls from the Cloverly Formation" (p. 79). These discoveries include a well-preserved skull of an Early Cretaceous triconodont. The alisphenoid in this animal was said to be comparatively larger than in *Morganucodon*, and contributes extensively to the structure of the side wall of the braincase (Clemens and Kielan-Jaworowska, 1979). Unfortunately, detailed study of the specimen awaits publication.

4. As far as multituberculates are concerned, Kermack misrepresented Simpson's (1937) data by saying: "In *Ptilodus* the nerve (V3) passes through what can only be the anterior lamina of the petrosal" (Kermack, 1967, p. 245). In fact, "Simpson (1937) avoided any statement implying a choice between these two bones for these foramina" (MacIntyre, 1967, p. 837). Clemens and Kielan-Jaworowska (1979, p. 109-110) admitted that "... recognition of the suture between the alisphenoid and petrosal is uncertain (see Kielan-Jaworowska, 1971, for further discussion). Identification of the large bone forming the lateral wall of the braincase in multituberculates as the ascending lamina of the petrosal was based in part on comparisons with the skull of *Ornithorhynchus*." Both Kermack and Kielan-Jaworowska *restored* the anterior lamina as based upon extrapolation from the skull of the platypus; each then used their reconstructions to suggest a close relatioship with monotremes. A perfect circle has been scribed!

5. As discussed earlier in this paper, in all known cases, morganucodontids appear to possess a large alisphenoid with an extensive ascending process (see Kermack and others, 1981, for *Morganucodon*; Crompton and Sun, 1985, for *Eozostrodon heikuopangensis*; and Gow, 1986, for *Megazostrodon*). However, in the case of *Morganucodon*, the gap between the alisphenoid and the anterior lamina as drawn by Kermack and others (1981) is an artifact. Therefore, one could equally argue that only a single bone was involved in that particular area of the braincase. In any rate, the real structure is equivocal.

If we are to admit that these are our actual database, should we have reason to document a nontherian/therian dichotomy based upon the difference in braincase pattern in the first place? Doubtless, the hypothesis should be abandoned. However, to philosophers and historians of science, this may represent a good example of how misunderstanding, misinterpretation, misrepresentation, circular reasoning, illogical deduction and/or induction, and even inaccurate observation can skew what we considered to be "knowledge" of the early evolution of mammals! This situation would be more easily understandable, if we remember: "No part of the aim of normal science is to call forth new sorts of phenomena; indeed those that will not fit the box are often not seen at all" (Kuhn, 1970, p. 24).

Phylogenetic Position of Multituberculata and Relationships of Early Mammals

Concerning the systematic position of Multituberculata, my discussion to this point is personally disappointing. However, if we accept the notion of monophyly of Mammalia, the new data derived from this study do provide important constraints to further formulation of hypotheses on relationships of early mammals, as well as for the phylogenetic position of Multituberculata.

First, I have thus far suggested that: (1) nontherians are not a natural assemblage; and (2) no compelling reason suggests a close relationship between multituberculates and monotremes. If these suggestions, as well as Archer and others' (1985) claim of *Steropodon* as a monotreme, are to be accepted, then therians' affinities with monotremes may become more credible (see also Kemp, 1983; Archer and others, 1985; and Kielan-Jaworowska and others, 1987). The present study also supports this assertion by having demonstrated that monotremes are more derived than multituberculates in having a much more reduced pila antotica, a curved cochlea, and undivided bony external nares. Therefore, mammalian phylogeny can be viewed as a tree, with monotremes and the other therians that shared the latest common ancestor with monotremes as the "crown" group. The rest of earlier mammalian groups (including multituberculates) are called the "stem" group, be they therians or nontherians (see Jefferies, 1979; C. Patterson, 1981; and Ax, 1985, 1987; for further discussions about stem versus crown group). Potentially, this approach may ease contentions involving phylogenetic analyses generated from different methodologies and/or different views in interpreting origin of certain characters. For example, in the three alternative cladograms that Miao and Lillegraven (1986, Fig. 2) presented, the different views about the origins of certain synapomorphies (such as the three ear ossicles and the reversed triangular pattern of molars) only affected the systematic positions of the members within the stem group. The systematic positions of the members within the crown group remained relatively stable. According to Jefferies (1979, p. 449), "Stem groups are extinct by definition and will be paraphyletic in that some members will be more closely related to the crown group than others are. All monophyletic groups containing more than one living species will, in principle, be divisible into a stem group and a crown group." Therefore, Multituberculata diverged from the main lineage leading to the crown group of Mammalia before emergence of the latest common ancestor of the crown group.

Second, within the stem group of Mammalia, multituberculates appear to be more derived than triconodonts, docodonts, and possibly kuehneotheriids in having evolved multiple cusp rows of molars and/or a triossicular middle ear. However, as Miao and Lillegraven (1986) documented, the relative position of multituberculates (compared to the other subgroups within the stem group) can be altered, depending upon different interpretations of the origins of certain attributes. Particularly, we then argued that discovery of three ear ossicles in *Lambdopsalis* (characteristic of modern mammals) was not adequate to determine whether the triossicular middle ear evolved once, twice, or multiply within the class Mammalia. In turn, the discovery did not help resolve the specific phylogenetic position of multituberculates. However, Allin (1986) and Sues (1986) have since shown that the configuration of a three-boned middle ear was probably already present in some advanced cynodonts. As Allin (1986, p. 292) argued, ". . . the commonality of antecedent morphology and function is so pervasive that a middle ear like that of mammals could have been expected to emerge in any cynodont derivative that survived long enough." Parallel origins of three mammalian middle ear ossicles, therefore, are explicitly suggested. In addition, Presley (*personal communication*) also suggested that the mammalian three ear ossicles could have been derived without further modification in any cynodont descendant that freed the post-dentary bones from the jaw (*i. e.,* in which the Meckel's cartilage atrophies anterior to the articular). The process, in fact, was observed in some marsupial foetals and newborns (Palmer, 1913; and Starck, 1967). Therefore, polyphyletic origin of the triossicular middle ear in mammals is not improbable. If indeed such were the case, phylogenetic relationships of early mammals as represented in the Figure 2B in Miao and Lillegraven (1986) would be favored.

Third, as stated at the beginning of this section, my discussion has been developed upon the assumption of monophyly of mammals. If, however, independent acquisition of the three-boned middle ear in multituberculates were probable, Simpson's (*e. g.,* 1928, 1929, 1937, 1945, 1971) life-long persistent opinion about the independent origin of multituberculates from cynodonts would have credibility. It is especially instructive, in my opinion, to read the following comment of Simpson's (1971, p. 191): "One could imagine multituberculate teeth evolving from those of morganucodonts or other triconodonts, but one could imagine almost anything, and in fact there is no evidence whatever that this occurred. Carrying the multituberculates back into Triassic haramiyids, itself highly dubious, does not help the case, because haramiyid teeth are likewise not at all like those of any triconodonts. In fact, they are rather more like the teeth of some therapsids." Admittedly, the phylogenetic position of Multituberculata within the class Mammalia may never be certain and, as a result, opinions on the issue may never become unified.

Finally, after all has been said, I am inclined to sustain our previous statement that, in terms of constraints provided by this study, ". . . a variety of previously held possibilities in reconstructing phylogenetic relationships among early groups of mammals can be reduced to only three. However, available paleontological evidence alone seems inadequate to allow favoritism for any one of the three alternatives" (Miao and Lillegraven, 1986, p. 500). As for other members in the stem group, multituberculates probably represented only part of the early radiation of mammals, and did not give rise to any of the three major groups of modern mammals; nor did multituberculates give rise to any of the fossil members of the crown group. Multituberculates diverged early from other mammals, a view shared by Kielan-Jawrowska and others (1986).

The study also provides a good example that illustrates the significance of fossils to determination of evolutionary relationships. But my frustration stands as testimony to what Colin Patterson (1981), himself a paleontologist, had to say about the place of paleontology in evolutionary biology: "What remains is the unity of the comparative method, in which paleontology can hold its own by acknowledging its debt to neontology, and by repaying that debt in contributing what it alone can: age of groups, paleobiogeographic data, and new character combinations that can reverse decisions on homology and polarity, so testing, and perhaps on rare occasions overthrowing, theories of relationship" (p. 220). I vigorously intended this study to carry out the spirit of that quotation.

CONCLUSIONS

Skull of *Lambdopsalis bulla* shows morphological unity with multituberculates in general, and with taeniolabidids in particular. The bulla-like structure of *Lambdopsalis* is the expanded vestibular apparatus of the inner ear, contrary to previous suggestions of its being either a tympanic bulla (Chow and Qi, 1978) or an air-filled, inflated tabular, paroccipital process, and mastoid (Kielan-Jaworowska and Sloan, 1979; and Kielan-Jaworowska and others, 1986). Lambdopsalidae Chow and Qi, 1978 was established as a monotypic family solely on the basis of the supposed presence of tympanic bulla in *Lambdopsalis*. In light of new information of cranial morphology (this study) and from the dental anatomy (Miao, 1986), *Lambdopsalis* is referred to Taeniolabididae Granger and Simpson, 1929. I consider Lambdopsalidae to be invalid, a subjective junior synonym of Taeniolabididae.

The extraordinary expansion of vestibular apparatus, flat incudomalleal joint, and absence of a well-defined fossa muscularis minor in *Lambdopsalis* suggest that *Lambdopsalis* was a burrower. Similar adaptations also may have occurred in ?*Catopsalis joyneri*. Hence, the presumed homogeneity in arboreal specializations of multituberculates may not reflect the diversity that these animals actually enjoyed, both in habits and in habitats.

Lambdopsalis possesses a typical mammalian palatal region. A reptilian ectopterygoid is not retained in *Lambdopsalis*, and its presumed retention in other multituberculates is questioned.

The alisphenoid (perforated by the trigeminal foramina) makes a major contribution to the side wall of the braincase in *Lambdopsalis*. The anterior lamina of the petrosal is represented only by a slender bone between the trigeminal and the prootic foramina in *Lambdopsalis*. This discovery strongly supports Presley's (1981) argument of close affinity between "nontherian" and "therian" mammals. It also exhibits the diversity of ossification of bony elements in the orbitotemporal region allowed by developmental processes, and the dangers of using this region for rigid phylogenetic interpretations. In addition, available paleontological evidences indicate nonexistence of a uniform structural pattern of the braincase for various so-called nontherian or "prototherian" groups. Thus the concept of a nontherian/therian dichotomy is invalid in terms of phylogeny or taxonomy.

A number of mammalian cranial characters that have been relied upon heavily in phylogenetic reconstructions are shown to be unreliable. These include premaxillo-frontal contact, exclusion of septomaxilla from the face, number of infraorbital foramina, extent of orbital exposure of palatine, presence versus absence of jugal, lacrimal, and parasphenoid, extent of the cranial process of squamosal, and reduction of alisphenoid. There is no *a priori* reason to believe that cranial characters are always useful or especially reliable. Therefore, character assessment is vital in phylogenetic analyses, and comparative methods remain powerful in detecting the homoplasy of certain cranial characters.

Previously-held interpretations of polarities of certain mammalian cranial characters are critically examined. The validity of some is questioned, such as sequence of the phylogenetic developments of the maxillo- and ethmoturbinals. Some polarities are reversed, such as presence versus absence of the optic foramen. Contrary to traditional belief, presence of a separate optic foramen both in *Lambdopsalis* and eutherians is shown as a symplesiomorphy of Mammalia, rather than as a synapomorphy of eutherians or a homoplasy between *Lambdopsalis* and eutherians.

Although discovery of three middle ear ossicles of *Lambdopsalis* alone did not help resolve whether the triossicular chain evolved once or more than once, a balanced view of both the fossil record and developmental mechanisms suggests that polyphyletic origin of the triossicular middle ear within the class Mammalia is a distinct possibility.

Cranial features coupled with dental characters suggest that nonharamiyid multituberculates form a monophyletic group. Nevertheless, more intimate relationships between multituberculates and any other major groups of mammals remain uncertain. Information derived from this study, however, provides important constraints to further formulation of phylogenetic hypotheses of mammals.

Assuming a monophyletic origin of mammals, the class Mammalia is divisible into a stem group and a crown group. The crown group includes all three major living groups of mammals (monotremes, marsupials, and eutherians) plus the fossil therians that shared the latest common ancestor with monotremes. The stem group consists of all remaining extinct mammals, be they therians or nontherians. Multituberculates are members of the paraphyletic stem group, and diverged from the main lineage leading to modern mammals prior to emergence of the latest common ancestor of modern mammals. Therefore, multituberculates have no particularly close relationship with any of the three major groups of living mammals, nor with any of the fossil members of the crown group.

Relationships among members of the stem group are difficult to ascertain, and depend largely upon interpretations of acquisitions of some key characters (such as triossicular middle ear and reversed triangular pattern of main molar cusps). However, in light of this study, a variety of previously held possibilities in reconstructing phylogenetic relationships among members of the stem group can be reduced to but three reasonable alternatives.

ACKNOWLEDGMENTS

I thank W. A. Clemens, Z. Kielan-Jaworowska, J. A. Lillegraven, and M. C. McKenna for suggesting this study as my PhD dissertation project, and for their unbridled support and enthusiasm throughout the course of the project. I also thank the members of my PhD Committee: D. W. Boyd, D. Duvall, D. M. Fountain, J. A. Lillegraven (Chairman), M. C. McKenna, and S. B. Smithson, for their services, suggestions, and support. Besides, it is my great pleasure to have the opportunity to know them as real gentlemen. I am especially indebted to my mentor, J. A. Lillegraven, for all that he has done for me during my four-year association with him. Even at the risk of being accused of distasteful applepolishing, I should add that it is his devotion to science and his commitment to vertebrate paleontology that have driven me to work, if ever, like a slave. Without his constant encouragement and unflagging support, I would not have finished this project. I only regret that I could not produce the kind of result that would fully repay his invaluable advice. My appreciation also is extended to L. E. Lillegraven for her many kindnesses.

At The University of Wyoming, I am also grateful to R. S. Houston (former Head of the Department of Geology and Geophysics, and Provost of the University), and J. E. McClurg (Head of the Department) for their encouragement and support. I also thank B. H. Breithaupt, K. M. Flanagan, W. A. Gern, Wendy Ware, Nancy Fallas, and Buff Moore, for their help and kindnesses.

At the IVPP, I thank Chow Minchen and Zhai Renjie for their constant support, encouragement, and help in all aspects before and throughout this study. I also thank Chow Minchen, Zhai Renjie, and especially Qi Tao for loan of the specimens. I pay tribute to Chang Yu-pin in memory of her great devotion to Chinese vertebrate paleontology and many kindnesses to me.

At the University of California, Berkeley, I thank W. A. Clemens, D. E. Savage, K. Padian, J. H. Hutchison, G. V. Shkurkin, J. A. Bacskai, L. J. Bryant, Luo Zhexi, and many graduate students at Department of Paleontology, who made my first-year stay in this friendly (though a bit strange at first glance) land enjoyable and memorable.

I also thank the following individuals and institutions for giving me access to comparative collections, and for their warm hospitalities: M. C. McKenna, M. J. Novacek, and R. H. Tedford, The American Museum of Natural History; J. J. Hooker, and A. W. Gentry, The British Museum (Natural History); J. A. Hopson, J. M.

Clark, and J. R. Wible, University of Chicago; B. Krebs, G. Krusat, and W. G. Kühne, Free University, Berlin; R. Presley, University College, Cardiff; T. S. Kemp, D. B. Norman, G. M. King, Oxford University; Z. Kielan-Jaworowska, Institute of Paleobiology, The Polish Academy of Sciences, Warsaw.

Informative discussions on various aspects of the study were provided by E. F. Allin, W. A. Clemens, J. A. Hopson, Z. Kielan-Jaworowska, J. A. Lillegraven, Luo Zhexi, M. C. McKenna, R. D. E. MacPhee, M. J. Novacek, R. Presley, and J. R. Wible. I thank J. A. Hopson, J. A. Lillegraven, M. C. McKenna, and J. R. Wible for critical reading of the entire manuscript, and for their invaluable comments and suggestions.

Financial support for this study was provided by teaching and research assistantships, fellowships (Graduate Education Funds), and scholarships (primarily from Hill Foundation) through the Department of Geology and Geophysics, The University of Wyoming, and through National Science Foundation Grant EAR 82-05211 and its continuation awarded to J. A. Lillegraven. International travel for the research was made possible through the generosity of Malcolm and Priscilla McKenna. Additional funds for the travels were provided by the Department of Geology and Geophysics, The University of Wyoming. Parts of the illustrations were made possible through NSF grant BSR 86-15016 to J. A. Hopson, and were done by Ms. C. Vanderslice (she also called attention to embarrassing neglect in my original descriptions). All of these are gratefully acknowledged.

Finally, many thanks to my mother, who saved my life during the great famine of the early 60s, and shows unflagging enthusiasm for my advanced education. And, against Chinese tradition, to my wife, Meng Xianna, for all her sacrifice.

REFERENCES CITED

Allin, E. F., 1975, Evolution of the mammalian middle ear: Journal of Morphology, v. 147, p. 403-438.

______ 1986, The auditory apparatus of advanced mammal-like reptiles and early mammals, *in* N. Hotton III, P. D. MacLean, J. J. Roth, and E. C. Roth, eds., The ecology and biology of mammal-like reptiles: Washington, Smithsonian Institution Press, p. 283-294.

Allison, A. C., 1953, The morphology of the olfactory system in the vertebrates: Biological Reviews of the Cambridge Philosophical Society, v. 28, p. 195-244.

Anderson, S., and Jones, J. K., Jr., 1984, Introduction, *in* S. Anderson, and J. K. Jones, Jr., eds., Orders and families of recent mammals of the world: New York, John Wiley & Sons, p. 1-10.

Archer, M., 1976, The basicranial region of marsupicarnivores (Marsupialia), interrelationships of carnivorous marsupials, and affinities of the insectivorous marsupial peramelids: Zoological Journal of the Linnean Society, v. 59, p. 217-322.

Archer, M., Flannery, T. F., Ritchie, A., and Molnar, R. E., 1985, First Mesozoic mammal from Australia—an early Cretaceous monotreme: Nature, v. 318, p. 363-366.

Archibald, J. D., 1977, Ectotympanic bone and internal carotid circulation of eutherians in reference to anthropoid origins: Journal of Human Evolution, v. 6, p. 609-622.

______ 1982, A study of Mammalia and geology across the Cretaceous-Tertiary boundary in Garfield County, Montana: University of California, Publications in Geological Sciences, v. 122, p. 1-286.

Ax, P., 1985, Stem species and the stem lineage concept: Cladistics, v. 1, p. 279-287.

______ 1987, The phylogenetic system: the systematization of organisms on the basis of their phylogenesis: New York, John Wiley & Sons Ltd., *xiii* + 340 p.

Barghusen, H. R., 1986, On the evolutionary origin of the therian tensor veli palatini and tensor tympani muscles, *in* N. Hotton III, P. D. MacLean, J. J. Roth, and E. C. Roth, eds., The ecology and biology of mammal-like reptiles: Washington, Smithsonian Institution Press, p. 253-262.

Bast, T. H., and Anson, B. J., 1949, The temporal bone and the ear: Springfield, Charles C. Thomas, Publisher, 478 p.

Bellairs, A. d'A., and Kamal, A. M., 1981, The chondrocranium and the development of the skull in recent reptiles, *in* C. Gans, and T. S. Parsons, eds., Biology of the Reptilia: v. 11, p. 1-263.

Bennett, A. F., and Ruben, J. A., 1986, The metabolic and thermoregulatory status of therapsids, *in* N. Hotton III, P. D. MacLean, J. J. Roth, and E. C. Roth, eds., The ecology and biology of mammal-like reptiles: Washington, Smithsonian Institution Press, p. 207-218.

Bonaparte, J. F., 1986, History of the terrestrial Cretaceous vertebrates of Gondwana: Actas IV Congreso Argentino de Paleontologia y Bioestratigrafia, Mendoza, Argentina, v. 2, p. 63-95.

Bramble, D. M., 1982, *Scaptochelys*: generic revision and evolution of gopher tortoises: Copeia, v. 4, p. 852-867.

Brink, A. S., 1957, Speculations on some advanced mammalian characteristics in the higher mammal-like reptiles: Palaeontologia Africana, v. 4, p. 77-96.

______ 1960, On some small therocephalians: *ibid.*, v. 7, p. 155-182.

______ 1980, The road to endothermy—a review: Societe Geologique de France, Memoires, New Series, n. 139, p. 29-38.

Broom, R., 1907, On the homology of the mammalian alisphenoid bone: South African Association for the advancement of Science, Report, 1907, p. 114-115.

______ 1909, Observations on the development of the marsupial skull: Linnean Society of New South Wales, Proceedings, v. 34, p. 24-212.

______ 1911, On the structure of the skull in the cynodont reptiles: Zoological Society of London, Proceedings, 1911, p. 893-925.

______ 1914, On the structure and affinities of the Multituberculata: American Museum of Natural History, Bulletin, v. 33, p. 115-134.

______ 1926, On the mammalian presphenoid and mesethmoid bones: Zoological Society of London, Proceedings, 1926, p. 257-264.

______ 1927, Some further points on the structure of the mammalian basicranial axis: Zoological Society of London, Proceedings, 1927, p. 233-244.

______ 1935, A further contribution to our knowledge of the structure of the mammalian basicranial axis: Transvaal Museum, Annals, v. 18, p. 33-36.

Bruner, H. L., 1907, On the cephalic veins and sinuses of reptiles, with a description of a mechanism for raising the venous blood pressure in the head: American Journal of Anatomy, v. 7, p. 1-117.

Bugge, J., 1972, The cephalic arterial system in the insectivores and the primates with special reference to the Macroscelidoidea and Tupaioidea and the insectivore-primate boundary: Zeitschrift fur Anatomie und Entwicklungsgeschichte, v. 135, p. 279-300.

______ 1974, The cephalic arterial system in insectivores, primates, rodents and lagomorphs, with special reference to the systematic classification: Acta Anatomica, v. 87, suppl. 62, p. 1-160.

______ 1985, Systematic value of the carotid arterial pattern in rodents, *in* W. P. Luckett, and J-L. Hartenberger, eds., Evolutionary relationships among rodents: a multidisciplinary analysis: NATO Series A: Life Sciences, v. 92, New York, Plenum Press, p. 355-379.

Butler, H., 1957, The development of certain human dural venous sinuses: Journal of Anatomy, v. 91, p. 510-526.

______ 1967, The development of mammalian dural venous sinuses with especial reference to the postglenoid vein: Journal of Anatomy, v. 102, p. 33-56.

Cartmill, M., and MacPhee, R. D. E., 1980, Tupaiid affinities: the evidence of the carotid arteries and cranial skeleton, *in* W. P. Luckett, ed., Comparative biology and evolutionary relationships of tree shrews: New York, Plenum Press, p. 95-132.

Chow, M., and Qi, T., 1978, Paleocene mammalian fossils from Nomogen Formation of Inner Mongolia: Vertebrata PalAsiatica, v. 16, p. 77-85 (in Chinese, with English summary).

Chow, M., Qi, T., and Li, Y., 1976, Paleocene stratigraphy and faunal characters of mammalian fossils in Nomogen Commune, Siziwang Qi County, Inner Mongolia: Vertebrata PalAsiatica, v. 14, p. 228-233 (in Chinese, with English summary).

Clark, J. M., and Hopson, J. A., 1985, Distinctive mammal-like reptile from Mexico and its bearing on the phylogeny of the Tritylodontidae: Nature, v. 315, p. 398-400.

Clemens, W. A., Jr., 1963, Fossil mammals of the type Lance Formation, Wyoming, Part I. Introduction and Multituberculata: University of California, Publications in Geological Sciences, v. 48, p. 1-105.

______ 1973, Fossil mammals of the type Lance Formation, Wyoming, Part III. Eutheria and summary: *ibid.*, v. 94, p. 1-102.

______ 1979, Notes on the Monotremata, *in* J. A. Lillegraven, Z. Kielan-Jaworowska, and W. A. Clemens, eds., Mesozoic mammals: the first two-thirds of mammalian history: Berkeley, University of California Press, p. 309-311.

Clemens, W. A., and Kielan-Jaworowska, Z., 1979, Multituberculata, *in ibid.*, p. 99-149.

Conroy, G. C., and Wible, J. R., 1978, Middle ear morphology of *Lemur variegatus*: some implications for primate paleontology: Folia Primatologica, v. 29, p. 81-85.

Cooper, C. F., 1928, On the ear region of certain of the Chrysochloridae: Royal Society of London, Philosophical Transactions (B), v. 216, p. 265-283.

Cope, E. D., 1881, On some Mammalia of the lowest Eocene beds of New Mexico: American Philosophical Society, Proceedings, v. 19, p. 484-495.

______ 1882, Mammalia in the Laramie Formation: American Naturalist, v. 16, p. 830-831.

______ 1884, The Tertiary *Marsupialia*: *ibid.*, v. 18, p. 686-697.

______ 1888, The Multituberculata monotremes: *ibid.*, v. 22, p. 259.

Cox, C. B., 1959, On the anatomy of a new dicynodont genus with evidence of the position of the tympanum: Zoological Society of London, Proceedings, v. 132, p. 321-367.

Crompton, A. W., 1958, The cranial morphology of a new genus and species of ictidosauran: *ibid.*, v. 130, p. 183-215.

______ 1964, On the skull of *Oligokyphus*: British Museum (Natural History), Bulletin, (Geology), v. 9, p. 69-82.

Crompton, A. W., and Jenkins, F. A., Jr., 1979, Origin of mammals, *in* J. A. Lillegraven, Z. Kielan-Jaworowska, and W. A. Clemens, eds., Mesozoic mammals: the first two-thirds of mammalian history: Berkeley, University of California Press, p. 59-73.

Crompton, A. W., and Sun, A., 1985, Cranial structure and relationships of the Liassic mammal *Sinoconodon*: Zoological Journal of the Linnean Society, v. 85, p. 99-119.

Darwin, C., 1896, The descent of man and selection in relation to sex: New York, D. Appleton and Company, 688 p.

De Beer, G. R., 1926, Studies on the vertebrate head, II. The orbito-temporal region of the skull: Quarterly Journal of Microscopic Science, v. 70, p. 263-370.

______ 1929, The development of the skull of the shrew: Royal Society of London, Philosophical Transactions (B), v. 217, p. 411-480.

______ 1937, The development of the vertebrate skull: Oxford, Clarendon Press, 552 p.

De Beer, G. R., and Fell, W. A., 1936, The development of the Monotremata.—Part III. The development of the skull of *Ornithorhynchus*: Zoological Society of London, Transactions, v. 23, p. 1-43.

Duvall, D., 1986, A new question of pheromones: aspects of possible chemical signaling and reception in the mammal-like reptiles, *in* N. Hotton III, P. D MacLean, J. J. Roth, and E. C. Roth, eds, The ecology and biology of mammal-like reptiles: Washington, Smithsonian Institution Press, p. 219-238.

Duvall, D., King, M. B., and Graves, B. M., 1983, Fossil and comparative evidence for possible chemical signaling in the mammal-like reptiles, *in* R. M. Silverstein, and D. Muller-Schwarze, eds., Chemical signals in vertebrates III: New York, Plenum Press, p. 25-44.

Edinger, T., and Kitts, D. B., 1954, The foramen ovale: Evolution, v. 8, p. 389-404.

Eldredge, N., and Cracraft, J., 1980, Phylogenetic patterns and the evolutionary process: method and theory in comparative biology: New York, Columbia University Press, 349 p.

Falconer, H., 1857, Description of two species of the fossil mammalian genus *Plagiaulax* from Purbeck: Geological Society of London, Quarterly Journal, v. 13, p. 261-282.

Fay, R. R., and Popper, A. N., 1985, The octavolateralis system, *in* M. Hildebrand, D. M. Bramble, K. F. Liem, and D. B. Wake, eds., Functional vertebrate morphology: Cambridge, Belknap Press, p. 291-316.

Fernandez, C., and Schmidt, R. S., 1963, The opossum ear and evolution of the coiled cochlea: Journal of Comparative Neurology, v. 121, p. 151-159.

Fleischer, G., 1978, Evolutionary principles of the mammalian middle ear: Advances in Anatomy, Embryology and Cell Biology, v. 55, p. 1-70.

Fourie, S., 1974, The cranial morphology of *Thrinaxodon liorhinus* Seeley: South African Museum, Annals, v. 65, p. 337-400.

Fuchs, H., 1908, Ueber einen Rest des Parasphenoids bei einem rezenten Säugertiere: Anatomischer Anzeiger, v. 32, p. 584.

——— 1911, Ueber das Septomaxillare eines rezenten Säugertiere (*Dasypus*), nebst einigen vergleichend-anatomischen Bemerkungen über das Septomaxillare und Praemaxillare der Amnioten überhaupt: *ibid.*, v. 38, p. 33-55.

——— 1915, Über den Bau und die Entwicklung des Schädels der *Chelone imbricata*. Ein Beitrag zur Entwicklungsgeschichte und vergleichenden Anatomie des Wirbeltierschädels. Erster Teil: Das Primordialskelett des Neurocraniums und des Kieferbogens, *in* A. Voeltzkow, ed., Reise in Ostafrika in den Jahren 1903-1905, wissenschaftliche Ergebnisse (v. 5): Stuttgart, Schweizerbart'sche Verlagsbuchhandlung, p. 1-325.

Gaughran, G. R. L., 1954, A comparative study of the osteology and myology of the cranial and cervical regions of the shrew, *Blarina brevicauda*, and the mole, *Scalopus aquaticus*: University of Michigan, Museum of Zoology, Miscellaneous Publications, no. 80, p. 1-82.

Gaupp, E., 1900, Das Chondrocranium von *Lacerta agilis*: Anatomische Hefte, Abt. 15, p. 433-595.

——— 1901, Alte Probleme und neuere Arbeiten über den Wirbeltierschädel: Ergebnisse der Anatomie und Entwickelungsgeschichte, v. 10, p. 847-1001.

——— 1902, Über die Ala temporalis des Säugerschädels und die Regio orbitalis einiger anderer Wirbeltierschädel: Anatomische Hefte, Abt. 19, p. 155-230.

——— 1904, A. Ecker's und R. Wiedersheim's Anatomie des Frosches: Braunschweig, Vieweg u. Sohn, v. 3, p. 1-961.

——— 1908, Zur Entwickelungsgeschichte und vergleichenden Morphologie des Schadels von *Echidna aculeata* var. typica: Denkschriften der Medizinisch-Naturwissenshaftlichen Gesellschaft zu Jena, v. 6, p. 539-788.

Ghiselin, M. T., 1976, The nomenclature of correspondence: a new look at "homology" and "analogy," *in* R. B. Masterson, W. Hodos, and H. Jerison, eds., Evolution, brain and behavior: persistent problems: Hillsdale, Lawrence Erlbaum, p. 129-142.

Gidley, J. W., 1909, Notes on the fossil mammalian genus *Ptilodus*, with descriptions of new species: U. S. National Museum, Proceedings, v. 36, p. 611-626.

Gilbert, G. K., 1896, The origin of hypotheses, illustrated by the discussion of a topographic problem: Science, v. 3, p. 1-13.

Gingerich, P. D., 1976, Cranial anatomy and evolution of early Tertiary Plesiadapidae (Mammalia, Primates): University of Michigan, Museum of Paleontology, Papers on Paleontology, no. 15, p. 1-141.

Goodrich, E. S., 1930, Studies on the structure and development of vertebrates: London, Macmillan, 837 p.

Gow, C. E., 1985, The side wall of the braincase in cynodont therapsids, and a note on the homology of the mammalian promontorium: South African Journal of Zoology, v. 21, p. 136-148.

——— 1986, A new skull of *Megazostrodon* (Mammalia, Triconodonta) from the Elliot Formation (Lower Jurassic) of southern Africa: Palaeontologia Africana, v. 26, p. 13-23.

Granger, W., and Simpson, G. G., 1929, A revision of the Tertiary Multituberculata: American Museum of Natural History, Bulletin, v. 56, p. 601-676.

Gray, O., 1955, A brief survey of the phylogenesis of the labyrinth: Journal of Laryngology and Otology, v. 69, p. 151-178.

Green, H. L. H. H., and Presley, R., 1978, The dumbbell bone of *Ornithorynchus*: Journal of Anatomy, v. 127, p. 216.

Gregory, W. K., 1910, The orders of mammals: American Museum of Natural History, Bulletin, v. 27, p. 1-524.

——— 1920, Studies in comparative myology and osteology: No. IV.—A review of the evolution of the lacrymal bone of vertebrates with special reference to that of mammals: American Museum of Natural History, Bulletin, v. 42, p. 95-263.

Gregory, W. K., and Noble, G. K., 1924, The origin of mammalian alisphenoid bone: Journal of Morphology and Physiology, v. 39, p. 435-463.

Griffiths, M., 1968, Echidnas: Oxford, Pergamon Press, 282 p.

______ 1978, The biology of the monotremes: New York, Academic Press, 367 p.

Guild, S. R., 1927, Observations upon the structure and normal contents of the ductus and saccus endolymphatics in the guinea pig (*Cavia cobaya*): American Journal of Anatomy, v. 39, p. 1-65.

Hahn, G., 1969, Beiträge zur Fauna der Grube Guimarota Nr. 3, die Multituberculata: Palaeontographica, Abt. A, v. 133, p. 1-100.

______ 1973, Neue Zähne von Haramiyiden aus der Deutschen Ober-Trias und ihre Beziehungen zu den Multituberculaten: *ibid.*, v. 142, p. 1-15.

______ 1977, Neue Schädel-Reste von Multituberculaten (Mamm.) aus dem Malm Portugals: Geologica et Palaeontologica, v. 11, p. 161-186.

______ 1978, Neue Unterkiefer von Multituberculaten aus dem Malm Portugals: *ibid.*, v. 12, p. 177-212.

______ 1981, Zum Bau der Schädel-Basis bei den Paulchoffatiidae (Multituberculata; Ober-Jura): Senckenbergiana Lethaea, v. 61, p. 227-245.

______ 1983, Überblick über die Erforschungs-Geschichte der Multituberculaten (Mammalia): Schriftenreihe fuer Geologische Wissenschaften, v. 19/20, p. 217-246.

______ 1985, Zum Bau des Infraorbital-Foramens bei den Paulchoffatiidae (Multituberculata, Ober-Jura): Berliner Geowissenschaften Abhandlung (A), v. 60, p. 5-27.

Hahn, G., and Hahn, R., 1983, Multituberculata, *in* F. Westphal, ed., Fossilum catalogus I: Animalia, pars 127: Amsterdam, Kugler Publications, p. 1-409.

Henson, O. W., Jr., 1961, Some morphological and functional aspects of certain structures of the middle ear in bats and insectivores: University of Kansas, Science Bulletin, v. 42, p. 151-255.

Hildebrand, M., 1974, Analysis of vertebrate structure: New York, John Wiley & Sons, 710 p.

Hill, J. E., 1935, The cranial foramina in rodents: Journal of Mammalogy, v. 16, p. 121-129.

Hinchcliffe, R., and Pye, A., 1969, Variations in the middle ear of the Mammalia: Journal of Zoology, London, V. 157, p. 277-288.

Hogben, L. T., 1919, The progressive reduction of the jugal in the Mammalia: Zoological Society of London, Proceedings, 1919, p. 71-78.

Hopson, J. A., 1964, The braincase of the advanced mammal-like reptile *Bienotherium*: Postilla, no. 87, p. 1-30.

______ 1969, The origin and adaptive radiation of mammal-like reptiles and nontherian mammals: New York Academy of Sciences, Annals, v. 167, p. 199-216.

______ 1970, The classification of nontherian mammals: Journal of Mammalogy, v. 51, p. 1-9.

Hopson, J. A., and Crompton, A. W., 1969, Origin of mammals, *in* T. Dobzhansky, M. K. Hecht, and W. C. Steere, eds., Evolutionary biology: v. 3, p. 15-72.

Huber, E., 1934, Anatomical notes on Pinnipedia and Cetacea: Carnegie Institution of Washington, Contributions to Palaeontology, no. 447, pt. 4, p. 106-136.

James, T. M., Presley, R., and Steel, F. L. D., 1980, The foramen ovale and sphenoidal angle in man: Anatomy and Embryology, v. 160, p. 93-104.

Jefferies, R. P. S., 1979, The origin of chordates—a methodological essay, *in* M. R. House, ed., The origin of major invertebrate groups: London, Academic Press, p. 443-477.

Jenkins, F. A., Jr., and Crompton, A. W., 1979, Triconodonta, *in* J. A. Lillegraven, Z. Kielan-Jaworowska, and W. A. Clemens, eds., Mesozoic mammals: the first two-thirds of mammalian history: Berkeley, University of California Press, p. 74-90.

Jenkins, F. A., Jr., and Krause, D. W., 1983, Adaptations for climbing in North American multituberculates (Mammalia): Science, v. 220, p. 712-715.

Jepsen, G. L., 1940, Paleocene faunas of the Polecat Bench Formation, Park County, Wyoming: American Philosophical Society, Proceedings, v. 83, p. 217-341.

Jollie, M., 1962, Chordate morphology: New York, Reinhold Publishing Corporation, 478 p.

______ 1968, The head skeleton of a new-born *Manis javanica* with comments on the ontogeny and phylogeny of the mammal head skeleton: Acta Zoologica (Stockholm), v. 49, p. 227-305.

Kemp, T. S., 1969, On the functional morphology of the gorgonopsid skull: Royal Society of London, Philosophical Transactions (B): v. 256, p. 1-83.

______ 1979, The primitive cynodont *Procynosuchus*: fuctional anatomy of the skull and relationships: *ibid.*, v. 285, p. 73-122.

______ 1980, Aspects of the structure and functional anatomy of the Middle Triassic cynodont *Luangwa*: Journal of Zoology, London, v. 191, p. 193-239.

______ 1982, Mammal-like reptiles and the origin of mammals: London, Academic Press, 363 p.

______ 1983, The relationships of mammals: Zoological Journal of the Linnean Society, v. 77, p. 353-384.

Kermack, D. M., and Kermack, K. A., 1984, The evolution of mammalian characters: Sydney, Croom Helm Ltd, 149 p.

Kermack, K. A., 1963, The cranial structure of the triconodonts: Royal Society of London, Philosophical Transactions (B): v. 246, p. 83-103.

______ 1967, The interrelations of early mammals: Journal of the Linnean Society (Zoology), v. 47, p. 241-249.

Kermack, K. A., and Kielan-Jaworowska, Z., 1971, Therian and non-therian mammals, *in* D. M. Kermack, and K. A. Kermack, eds., Early mammals: Zoological Journal of the Linnean Society, v. 50, suppl. 1, p. 103-115.

Kermack, K. A., and Mussett, F., 1958, The jaw articulation of the Docodonta and the classification of Mesozoic mammals: Royal Society of London, Proceedings (B), v. 149, p. 204-215.

Kermack, K. A., Mussett, F., and Rigney, H. W., 1973, The lower jaw of *Morganucodon*: Zoological Journal of the Linnean Society, v. 53, p. 87-175.

______ 1981, The skull of *Morganucodon*: *ibid.*, v. 71, p. 1-158.

Kesteven, H. L., 1918, The homology of the mammalian alisphenoid and of the echdina pterygoid: Journal of Anatomy, v. 52, p. 223-238.

Kielan-Jaworowska, Z., 1970, Unknown structures in multituberculate skull: Nature, v. 226, p. 974-976.

______ 1971, Skull structure and affinities of the Multituberculata, *in* Z. Kielan-Jaworowska, ed., Results of the Polish-Mongolian Palaeontological Expeditions, Part III: Palaeontologia Polonica, no. 25, p. 5-41.

______ 1974, Multituberculate succession in the Late Cretaceous of the Gobi Desert (Mongolia), *in ibid.*, Part V, no. 30, p. 23-44.

______ 1983, Multituberculate endocranial casts: Palaeovertebrata, v. 13, p. 1-12.

______ 1986, Brain evolution in Mesozoic mammals, *in* K. M. Flanagan, and J. A. Lillegraven, eds., Vertebrates, phylogeny, and philosophy: Contributions to Geology, University of Wyoming, Special Paper 3, p. 21-34.

Kielan-Jaworowska, Z., Crompton, A. W., and Jenkins, F. A., Jr., 1987, The origin of egg-laying mammals: Nature, v. 326, p. 871-873.

Kielan-Jaworowska, Z., and Dashzeveg, D., 1978, New Late Cretaceous mammal locality in Mongolia and a description of a new multituberculate: Acta Palaeontologica Polonica, v. 23, p. 115-130.

Kielan-Jaworowska, Z., Poplin, C., Presley, R., and Ricqles, R. & de, 1984, Preliminary data on multituberculate cranial anatomy studied by serial sections, *in* W. -E. Reif, and F. Westphal, eds., Third symposium on Mesozoic terrestrial ecosystem: Tubingen, Attempto Verlag, p. 123-128.

Kielan-Jaworowska, Z., Presley, R., and Poplin, C., 1986, The cranial vascular system in taeniolabidoid multituberculate mammals: Royal Society of London, Philosophical Transactions (B), V. 313, p. 525-602.

Kielan-Jaworowska, Z., and Sloan, R. E., 1979, *Catopsalis* (Multituberculata) from Asia and North America and the problem of taeniolabidid dispersal in the Late Cretaceous: Acta Palaeontologica Polonica, v. 24, p. 187-197.

Kingsley, J. S., 1926, Outlines of comparative anatomy of vertebrates (Third edition): Philadelphia, P. Blakiston's Son & Co., 470 p.

Kohncke, M., 1985, The chondrocranium of *Cryptoprocta ferox*: Advances in Anatomy, Embryology and Cell Biology, v. 95, p. 1-89.

Krause, D. W., 1986, Competitive exclusion and taxonomic displacement in the fossil record: the case of rodents and multituberculates in North America, *in* K. M. Flanagan, and J. A. Lillegraven, eds., Vertebrates, phylogeny, and philosophy: Contributions to Geology, University of Wyoming, Special Paper 3, p. 95-118.

Krause, D. W., and Jenkins, F. A., Jr., 1983, The postcranial skeleton of North American multituberculates: Harvard University, Museum of Comparative Zoology, Bulletin, v. 150, p. 199-246.

Kuhn, H.-J., and Zeller, U., 1987, The cavum epiptericum in monotremes and therian mammals, *in* H.-J. Kuhn, and U. Zeller, eds., Morphogenesis of the mammalian skull: Hamburg, Verlag Paul Parey, p. 51-70.

Kuhn, T. S., 1970, The structure of scientific revolutions (Second edition): Chicago, University of Chicago Press, 210 p.

Kühne, W. G., 1956, The Liassic therapsid *Oligokyphus*: London, British Museum (Natural History), 147 p.

Lay, D. M., 1972, The anatomy, physiology, functional significance and evolution of specialized hearing organ of gerbilline rodents: Journal of Morphology, v. 138, p. 41-120.

Lewis, E. R., Leverenz, E. L., and Bialek, W. S., 1985, The vertebrate inner ear: Boca Raton, CRC Press, Inc., 248 p.

Li, C., and Ting, S., 1983, The Paleogene mammals of China: Carnegie Museum of Natural History, Bulletin, no. 21, p. 1-93.

Lillegraven, J. A., and McKenna, M. C., 1986, Fossil mammals from the "Mesaverde" Formation (Late Cretaceous, Judithian) of the Bighorn and Wind River Basins, Wyoming, with definitions of Late Cretaceous North American Land-Mammal "Ages:" American Museum Novitates, no. 2840, p. 1-68.

Lorenz, K. Z., 1981, The foundations of ethology: New York, Simon and Schuster, 380 p.

MacIntyre, G. T., 1967, Foramen pseudovale and quasi-mammals: Evolution, v. 21, p. 834-841.

MacPhee, R. D. E., 1981, Auditory regions of primates and eutherian insectivores: morphology, ontogeny, and character analysis: Contributions to Primatology, v. 18, p. 1-282.

MacPhee, R. D. E., and Cartmill, M., 1986, Basicranial structures and primate systematics, *in* D. R. Swindler, ed., Comparative primate biology, v. 1: Systematics, evolution, and anatomy: New York, Alan R. Liss, Inc., p. 219-275.

McDowell, S. B., Jr., 1958, The greater antillean insectivores: American Museum of Natural History, Bulletin, v. 115, p. 113-214.

McKenna, M. C., 1976, Comments on Radinsky's "Later mammal radiations," *in* R. B. Masterson, M. E. Bitterman, C. B. G. Campbell, and N. Hotton III, eds., Evolution of brain and behavior in vertebrates, v. 1: Hillsdale, Lawrence Erlbaum Associates, Publishers, p. 245-250.

______ 1987, Molecular and morphological analysis of high-level mammalian interrelationships, *in* Patterson, C., ed., Molecules and morphology in evolution: conflict or compromise?: Cambridge, Cambridge University Press, p. 55-93.

McNab, B. K., 1978, The evolution of endothermy in the phylogeny of mammals: American Naturalist, v. 112, p. 1-21.

Maier, W., 1987, The ontogenetic development of the orbitotemporal region in the skull of *Monodelphis domestica* (Didelphidae, Marsupialia), and the problem of the mammalian alisphenoid, *in* H. -J. Kuhn, and U. Zeller, eds., Morphogenesis of the mammalian skull: Hamburg, Verlag Paul Parey, p. 71-90.

Marshall, L. G., 1979, Evolution of metatherian and eutherian (mammalian) characters: a review based on cladistic methodology: Zoological Journal of the Linnean Society, v. 66, p. 369-410.

Matthew, W. D., 1909, Carnivora and Insectivora of the Bridger Basin, Middle Eocene: American Museum of Natural History, Memoirs, v. 9, p. 289-567.

Matthew, W. D., and Granger, W., 1925, Fauna and correlation of the Gashato Formation of Mongolia: American Museum Novitates, no. 189, p. 1-12.

Maynard Smith, J., 1984, Palaeontology at the high table: Nature, v. 309, p. 401-402.

Miao, D., 1986, Dental anatomy and ontogeny of *Lambdopsalis bulla* (Mammalia, Multituberculata): Contributions to Geology, University of Wyoming, v. 24, p. 65-76.

Miao, D., and Lillegraven, J. A., 1986, Discovery of three ear ossicles in a multituberculate mammal: National Geographic Research, v. 2, p. 500-507.

Moore, W. J., 1981, The mammalian skull: Cambridge, Cambridge University Press, 369 p.

Muller, J., 1935, The orbitotemporal region of the skull of the Mammalia: Archives Neerlandaises de Zoologie, v. 1, p. 118-259.

Novacek, M. J., 1980, Cranioskeletal features in tupaiids and selected Eutheria as phylogenetic evidence, *in* W. P. Luckett, ed., Comparative biology and evolutionary relationships of tree shrews: New York, Plenum Press, p. 35-93.

______ 1985, Cranial evidence for rodent affinities, *in* W. P. Luckett, and J-L. Hartenberger, eds., Evolutionary relationships among rodents: a multidisciplinary analysis: NATO Series A: Life Sciences, v. 92: New York, Plenum Press, p. 59-81.

______ 1986, The skull of leptictid insectivorans and the higher-level classification of eutherian mammals: American Museum of Natural History, Bulletin, v. 183, p. 1-112.

Nowak, R. M., and Paradiso, J. L., 1983, Walker's mammals of the world (Fourth edition, 2 vols.): Baltimore, Johns Hopkins University Press, 1362 p.

O'Donoghue, C. H., The blood vascular system of the tuatara, *Sphenodon punctatus*: Royal Society of London, Philosophical Transactions (B), v. 210, p. 175-252.

Oelrich, T. M., 1956, The anatomy of the head of *Ctenosaura pectinata* (Iguanidae): University of Michigan, Museum of Zoology, Miscellaneous Publications, no. 94, p. 1-122.

Olson, E. C., 1944, Origin of mammals based upon cranial morphology of the therapsid suborders: Geological Society of America, Special Papers, no. 55, p. 1-136.

Olson, S. J., 1985, Origins of the domestic dog: the fossil record: Tucson, University of Arizona Press, 118 p.

Osborn, H. F., 1895, The heredity mechanism and the search for the unknown factors of evolution: American Naturalist, v. 29, p. 418-439.

Palmer, R. W., 1913, Note on the lower jaw and ear ossicles of a foetal *Perameles*: Anatomischer Anzeiger, v. 43, p. 510-515.

Parker, W. K., 1871, On the structure and development of the skull of the common frog: Royal Society of London, Philosophical Transactions, v. 161, p. 137-211.

______ 1885, On the structure and development of the skull in the Mammalia: *ibid.*, v. 176, pt. 2: Edentata, p. 1-120; pt. 3: Insectivora, p. 121-275.

Parker, W. N., 1897, Elements of the comparative anatomy of vertebrates (Second edition): London, Macmillan and Co., 488 p.

______ 1907, Comparative anatomy of vertebrates (Third edition): London, Macmillan and Co., 576 p.

Parrington, F. R., 1946, On the cranial anatomy of cynodonts: Zoological Society of London, Proceedings, v. 116, p. 181-197.

______ 1971, On the Upper Triassic mammals: Royal Society of London, Philosophical Transactions (B), v. 261, p. 231-272.

Parrington, F. R., and Westoll, T. S., 1940, On the evolution of the mammalian palate: *ibid.*, v. 230, p. 305-355.

Parsons, T. S., 1959, Nasal anatomy and the phylogeny of reptiles: Evolution, v. 13, p. 175-187.

______ 1971, Anatomy of nasal structures from a comparative viewpoint, *in* L. M. Beidler, ed., Handbook of sensory physiology, v. 4, Chemical senses, pt. 2, Olfaction: New York, Springer-Verlag.

Patterson, B., 1965, The auditory region of the borhyaenid marsupial *Cladosictis*: Breviora, no. 217, p. 1-9.

Patterson, B., and Olson, E. C., 1961, A triconodontid mammal from the Triassic of Yunnan: International Colloquium on the Evolution of Lower and Nonspecialized Mammals, pt. 1, p. 129-191.

Patterson, C., 1977, Cartilage bones, dermal bones and membrane bones, or the exoskeleton versus the endoskeleton, *in* S. M. Andrews, R. S. Miles, and A. D. Walker, eds., Problems in vertebrate evolution: New York, Academic Press, p. 77-121.

______ 1981, Significance of fossils in determining evolutionary relationships: Annual Review of Ecology and Systematics, v. 12, p. 195-223.

Paulli, S., 1900*a*, Über die Pneumaticitat des Schädels bei den Säugethieren, I. Über den Bau des Siebbeins. Über die Morphologie des Siebbeins und die der Pneumaticität bei den Monotremen und den Marsupialiern: Gegenbaurs Morphologisches Jahrbuch, v. 28, p. 147-178.

______ 1900*b*, Über die Pneumaticität des Schädels bei den Säugethieren, II. Über die Morphologie des Siebbeins und die Pneumaticität bei den Ungulaten und Probosciden: *ibid.*, v. 28, p. 179-251.

——— 1900*c*, Über die Pneumaticität des Schädels bei den Säugethieren, III. Über die Morphologie des Siebbeins und Pneumaticität bei den Insectivoren, Hyracoideen, Chiropteren, Carnivoren, Pinnipedien, Edentaten, Rodentiern, Prosimien und Primaten: *ibid.*, v. 28, p. 483-564.

Pickles, J. O., 1982, An introduction to the physiology of hearing: London, Academic Press, 341 p.

Presley, R., 1979, The primitive course of the internal carotid artery in mammals: Acta Anatomica, v. 103, p. 238-244.

——— 1980, The braincase in Recent and Mesozoic therapsids: Societe Geologique de France, Memoires, New Series, no. 139, p. 159-162.

——— 1981, Alisphenoid equivalents in placentals, marsupials, monotremes and fossils: Nature, v. 294, p. 668-670.

Presley, R., and Steel, F. L. D., 1976, On the homology of the alisphenoid: Journal of Anatomy, v. 121, p. 441-459.

——— 1978, The pterygoid and ectopterygoid in mammals: Anatomy and Embryology, v. 154, p. 95-110.

Pye, A., and Hinchcliffe, R., 1976, Comparative anatomy of the ear, *in* R. Hinchcliffe, and D. Harrison, eds., Scientific foundations of otolaryngology: Chicago, William Heinemann Medical Books Publication, p. 184-202.

Radinsky, L. B., 1964, Notes on Eocene and Oligocene fossil localities in Inner Mongolia: American Museum Novitates, no. 2180, p. 1-11.

Rajtova, V., 1972, Morphogenesis des Chondrocraniums beim Meerschweinchen (*Cavia porcellus*): Anatomischer Anzeiger, v. 130, p. 176-206.

Reighard, J., and Jennings, H. S., 1935, Anatomy of the cat (Third edition): New York, Henry Holt and Company, 486 p.

Reinbach, W., 1952*a*, Zur Entwicklung des Primordialcraniums von *Dasypus novemcinctus* Linné (*Tatusia novemcincta* Lessen) I.: Zeitschrift für Morphologie und Anthropologie, v. 44, p. 375-444.

——— 1952*b*, Zur Entwicklung des Primordialcraniums von *Dasypus novemcinctus* Linné (*Tatusia novemcincta* Lessen) II.: *ibid.*, v. 45, p. 1-72.

Ride, W. D. L., 1957, The affinities of *Plagiaulax* (Multituberculata): Zoological Society of London, Proceedings, v. 128, p. 397-402.

Romanes, G. J., ed., 1981, Cunningham's textbook of anatomy (Twelfth edition): Oxford, Oxford University Press, 1078 p.

Romer, A. S., 1956, Osteology of the reptiles: Chicago, University of Chicago Press, 772 p.

——— 1966, Vertebrate paleontology (Third edition): Chicago, University of Chicago Press, 468 p.

——— 1970, The Chanares (Argentina) Triassic reptile fauna VI, a chiniquodontid cynodon with an incipient squamosal-dentary jaw articulation: Breviora, no. 344, p. 1-18.

Romer, A. S., and Parsons, T. S., 1977, The vertebrate body (Fifth edition): Philadelphia, Saunders Company, 624 p.

Roux, G. H., 1947, The cranial development of certain Ethiopian "insectivores" and its bearing on the mutual affinities of the group: Acta Zoologica (Stockholm), v. 28, p. 165-397.

Rowe, T., 1986, Osteological diagnosis of Mammalia, L. 1758, and its relationship to extinct Synapsida: Ph.D. Dissertation, University of California at Berkeley, 446 p.

Rudwick, M. J. S., 1976, The meaning of fossils: episodes in the history of palaeontology (Second edition): New York, Neale Watson Academic Publications, Inc., 287 p.

Schoch, R. M., 1984, Introduction, *in* R. M. Schoch, ed., Vertebrate paleontology: New York, Van Nostrand Reinhold Company, p. 1-16.

Shindo, T., 1915, Bedeutung des Sinus cavernosus der Säuger mit vergleichend anatomischer Berücksichtigung anderer Kopfvenen: Anatomische Hefte, v. 52, p. 319-495.

Shoshani, J., 1986, Mammalian phylogeny: comparison of morphological and molecular results: Molecular Biology and Evolution, v. 3, p. 222-242.

Simmons, N. B., and Miao, D., 1986, Paraphyly in *Catopsalis* (Mammalia: Multituberculata) and its biogeographic implications, *in* K. M. Flanagan, and J. A. Lillegraven, eds., Vertebrates, phylogeny, and philosophy: Contributions to Geology, University of Wyoming, Special Paper 3, p. 87-94.

Simpson, G. G., 1925, A Mesozoic mammal skull from Mongolia: American Museum Novitates, no. 201, p. 1-11.

——— 1926, Mesozoic Mammalia. IV. The multituberculates as living animals: American Journal of Science, ser. 5, v. 11, p. 228-250.

——— 1928, A catalogue of the Mesozoic Mammalia in the Geological Department of the British Museum: London, Oxford University Press, 215 p.

——— 1929, The dentition of *Ornithorhynchus* as evidence of its affinities: American Museum Novitates, no. 390, p. 1-15.

——— 1937, Skull structure of the Multituberculata: American Museum of Natural History, Bulletin, v. 73, p. 727-763.

——— 1938, Osteography of the ear region in monotremes: American Museum Novitates, no. 978, p. 1-15.

——— 1945, The principles of classification and a classification of mammals: American Museum of Natural History, Bulletin, v. 85, p. 1-350.

——— 1970, The Argyrolagidae, extinct South American marsupials: Harvard University, Museum of Comparative Zoology, Bulletin, v. 139, p. 1-86.

——— 1971, Concluding remarks: Mesozoic mammals revisited, *in* D. M. Kermack, and K. A. Kermack, eds., Early mammals: Zoological Journal of the Linnean Society, v. 50, suppl. 1, p. 181-198.

Slade, D. D., 1895, The significance of the jugal arch: American Philosophical Society, Proceedings, v. 34, p. 50-67.

Sloan, R. E., 1979, Multituberculata, *in* R. W. Fairbridge, and D. Jablonski, eds., The encyclopedia of paleontology: Stroudsburg, Dowden, Hutchinson & Ross Inc., p. 492-498.

Smith, C. A., and Takasaka, T., 1971, Auditory receptor organs of reptiles, birds, and mammals, *in* W. D. Neff, ed., Contributions to sensory physiology, v. 5: New York, Academic Press, p. 129-178.

Starck, D., 1967, Le crâne des mammifères, *in* P-P. Grassé, ed., Traité de zoologie: anatomie, systematique, biologie: Paris, Masson, p. 405-549.

______ 1979, Cranio-cerebral relations in recent reptiles, *in* C. Gans, R. G. Northcutt, and P. Ulinski, eds., Biology of the Reptilia, v. 9: London, Academic Press, p. 1-38.

Story, H. E., 1951, The carotid arteries in the Procyonidae: Fieldiana (Zoology), v. 32, p. 477-557.

Struthers, P. H., 1927, The prenatal skull of the Canadian porcupine (*Erethizon dorsatus*): Journal of Morphology and Physiology, v. 44, p. 127-216.

Sues, H. -D., 1985, The relationships of the Tritylodontidae (Synapsida): Zoological Journal of the Linnean Society, v. 85, p. 205-217.

______ 1986, The skull and dentition of two tritylodontid synapsids from the Lower Jurassic of western North America: Harvard University, Museum of Comparative Zoology, Bulletin, v. 151, p. 217-268.

Sutton, J. B., 1888, A critical study in cranial morphology: Journal of Anatomy and Physiology, London, v. 22, p. 23-37.

Tatarinov, L. P., 1963, New Late Permian therocephalian: Palaeontologicheskiy Zhurnal, 1963, p. 76-94 (in Russian).

Terry, R. J., 1917, The primordial cranium of the cat: Journal of Morphology, v. 29, p. 281-433.

Thomason, J. J., and Russell, A. P., 1986, Mechanical factors in the evolution of the mammalian secondary palate: a theoretical analysis: *ibid.*, v. 189, p. 199-213.

Thyng, F. W., 1906, Squamosal bone in tetrapodous Vertebrata: Boston Society of Natural History, Proceedings, v. 32, p. 387.

Turnbull, W. D., 1970, Mammalian masticatory apparatus: Fieldiana (Geology), v. 18, p. 149-356.

Tyndale-Biscoe, H., and Renfree, M., 1987, Reproductive physiology of marsupials: Cambridge, Cambridge University Press, 476 p.

Vandebroek, G., 1964, Recherches sur l'origine des mammifères: Société Royale Zoologique de Belgique, Annales, v. 94, p. 117-160.

Van Valen, L., 1959, Therapsids as mammals: Evolution, v. 14, p. 304-313.

Van Valen, L., and Sloan, R. E., 1966, The extinction of the multituberculates: Systematic Zoology, v. 15, p. 261-278.

Wahlert, J. H., 1974, The cranial foramina of protrogomorphous rodents; an anatomical and phylogenetic study: Harvard University, Museum of Comparative Zoology, Bulletin, v. 146, p. 363-410.

______ 1977, Cranial foramina and relationships of *Eutypomys* (Rodentia, Eutypomyidae): American Museum Novitates, no. 2626, p. 1-8.

______ 1978, Cranial foramina and relationships of the Eomyoidea (Rodentia, Geomorpha). Skull and upper teeth of *Kansasimys*: *ibid.*, no. 2645, p. 1-16.

______ 1983, Relationships of the Florentiamyidae (Rodentia, Geomyoidea) based on cranial and dental morphology: *ibid.*, no. 2769, p. 1-23.

______ 1985, Cranial foramina of rodents, *in* W. P. Luckett, and J-L. Hartenberger, eds., Evolutionary relationships among rodents: a multidisciplinary analysis: NATO Series, ser. A: Life Sciences, v. 92: New York, Plenum Press, p. 311-332.

Walker, W. F., 1986, Vertebrate dissection (Seventh edition): Philadelphia, Saunders Company, 391 p.

Wang, J. C. C., and Lung, M. A. K. Y., 1984, Nasal blood flow in the dog, *in* S. Hunyor, J. Ludbrook, J. Shaw, and M. McGrath, eds., The peripheral circulation: New York, Elsevier Science Publishers, p. 149-151.

Watson, D. M. S., 1911, The skull of *Diademodon*, with notes on those of some other cynodonts: Annals and Magazine of Natural History, ser. 8, v. 8, p. 293-330.

______ 1913, Further notes on the skull, brain and organs of special sense of *Diademodon*: *ibid.*, ser. 8, v. 12, p. 217-228.

______ 1916, The monotreme skull: a contribution to mammalian morphogenesis: Royal Society of London, Philosophical Transactions (B), v. 207, p. 311-374.

______ 1920, On the Cynodontia: Annals and Magazine of Natural History, ser. 9, v. 6, p. 506-524.

______ 1951, Paleontology and modern biology: New Haven, Yale University Press, 216 p.

______ 1953, The evolution of the mammalian ear: Evolution, v. 7, p. 159-177.

Webster, D. B., 1961, The ear apparatus of the kangaroo rat, *Dipodomys*: American Journal of Anatomy, v. 108, p. 123-148.

______ 1962, A function of the enlarged middle-ear cavities of the kangaroo rat, Dipodomys: Physiological Zoology, v. 35, p. 248-255.

______ 1966, Ear structure and function in modern mammals: American Zoologist, v. 6, p. 451-466.

______ 1970, The cochlear nuclei of Heteromyidae: *ibid.*, v. 10, p. 554-555.

______ 1973, Audition, vision, and olfaction in kangaroo rat predator avoidance: *ibid.*, v. 13, p. 1346.

Webster, D. B., and Webster, M., 1971, Adaptive value of hearing and vision in kangaroo rat predator avoidance: Brain, Behavior, and Evolution, v. 4, p. 310-322.

______ 1972, Kangaroo rat auditory thresholds before and after middle ear reduction: *ibid.*, v. 5, p. 41-53.

______ 1975, Auditory systems of Heteromyidae: functional morphology and evolution of the middle ear: Journal of Morphology, v. 146, p. 343-376.

______ 1977, Auditory systems of Heteromyidae: cochlear diversity: *ibid.*, v. 152, p. 153-170.

——— 1980, Morphological adaptations of the ear in the rodent family Heteromyidae: American Zoologist, v. 20, p. 247-254.

——— 1984, The specialized auditory system of kangaroo rats, *in* W. D. Neff, ed., Contributions to sensory physiology, v. 8: New York, Academic Press, p. 161-196.

Webster, M., 1977, Kangaroo rat cochlea: qualitatively and quantitatively unique features: American Academy of Ophthalmology and Otolaryngology, Transactions, V. 84, p. 223-232.

Wegner, R. N., 1922, Der Stutzknochen, Os nariale, in der Nasenhohle bei den Gurteltieren, Dasypodidae, und seine homologen Gebilde bei Amphibien, Reptilien und Monotremen: Gegenbaurs Morphologische Jahrbuch, v. 51, p. 413-492.

Wever, E. G., 1978, The reptile ear: its structure and function: Princeton, Princeton University Press, 1024 p.

——— 1985, The amphibian ear: Princeton, Princeton University Press, 488 p.

Wheeler, Q. D., 1986, Character weighting and cladistic analysis: Systematic Zoology, v. 35, p. 102-109.

Wible, J. R., 1983, The internal carotid artery in early eutherians: Acta Palaeontologica Polonica, v. 28, p. 281-293.

——— 1984, The ontogeny and phylogeny of the mammalian cranial arterial pattern: Ph.D. Dissertation, Duke University, Durham, 705 p.

——— 1986, Transformations in the extracranial course of the internal carotid artery in mammalian phylogeny: Journal of Vertebrate Paleontology, v. 6, p. 313-325.

——— 1987, The eutherian stapedial artery: character analysis and implications for superordinal relationships: Zoological Journal of the Linnean Society, v. 91, p. 107-135.

——— *in press*, Vessels on the side wall of the braincase in cynodonts and primitive mammals: Fortschritte der Zoologie.

Winge, H., 1923, Pattedyr-Slaegter, v. 1: Copenhagen, 360 p.

Yapp, W. B., 1965, Vertebrates: their structure and life: New York, Oxford University Press, 525 p.

Young, C. C., 1947, Mammal-like reptiles from Lufeng, Yunnan, China: Zoological Society of London, Proceedings, v. 117, p. 537-596.

Young, E. D., Fernandez, C., and Goldberg, J. M., 1976, Sensitivity of vestibular nerve fibers to audiofrequency sound and head vibration in the squirrel monkey: Acoustic Society of America, Journal, v. 59, p. 47.

Zeller, U., 1986, The systematic relations of tree shrews: evidence from skull morphogenesis, *in* J. G. Else, and P. C. Lee, eds., Primate evolution, v. 1: London, Cambridge University Press, p. 273-280.